SCIENCE
UNIVERSAL LIBRARY

REPORT
ON THE
SCIENTIFIC RESULTS
OF THE
VOYAGE OF H.M.S. CHALLENGER
DURING THE YEARS 1873-76

COLOUR PICTURE

挑战者号航海考察科学成果报告

彩图集

[英] 约翰·默里 主编

华东师范大学出版社

图书在版编目(CIP)数据

挑战者号航海考察科学成果报告.彩图集 = The Report on the Scientific Results of the Voyage of H.M.S Challenger Colour Picture：英文 /（英）约翰·默里(John Murray) 主编. — 上海：华东师范大学出版社, 2018
 （寰宇文献）
 ISBN 978-7-5675-7325-3

Ⅰ.①挑… Ⅱ.①约… Ⅲ.①海洋-科学考察-世界-图集 Ⅳ.①P72

中国版本图书馆CIP数据核字(2018)第002081号

挑战者号航海考察科学成果报告 彩图集
The Report on the Scientific Results of the Voyage of H.M.S Challenger Colour Picture
（英）约翰·默里 (John Murray) 主编

特约策划	黄曙辉
责任编辑	庞　坚
特约编辑	许　倩
装帧设计	刘怡霖　崔　明

出版发行	华东师范大学出版社
社　　址	上海市中山北路3663号　邮编 200062
网　　址	www.ecnupress.com.cn
电　　话	021-60821666　行政传真　021-62572105
客服电话	021-62865537
门市（邮购）电话	021-62869887
地　　址	上海市中山北路3663号华东师范大学校内先锋路口
网　　店	http://hdsdcbs.tmall.com/

印 刷 者	虎彩印艺股份有限公司
开　　本	16开
印　　张	25.5
版　　次	2018年1月第1版
印　　次	2018年1月第1次
书　　号	ISBN 978-7-5675-7325-3
定　　价	320.00元

出 版 人　王　焰

（如发现本版图书有印订质量问题，请寄回本社客服中心调换或电话021-62865537联系）

第一卷动物志一彩图

A. HOLOCENTRUM SANCTI PAULI.
(I.st of St Paul.)
B. BATHYANTHIAS ROSEUS. C. CENTROPRISTIS ANNULARIS. D. HIPPOCAMPUS VILLOSUS.
(Coast of Brazil.)

A. PERISTETHUS TRUNCATUM. B. RHOMBOIDICHTHYS CORNUTUS.
(Pernambuco.)
C. SPINAX GRANULOSUS.
(Chile)

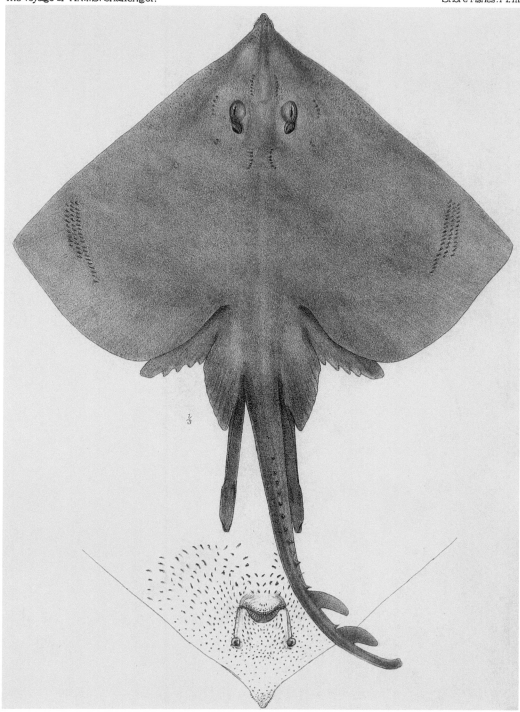

RAJA PLATANA.
(Mouth of the Rio de la Plata.)

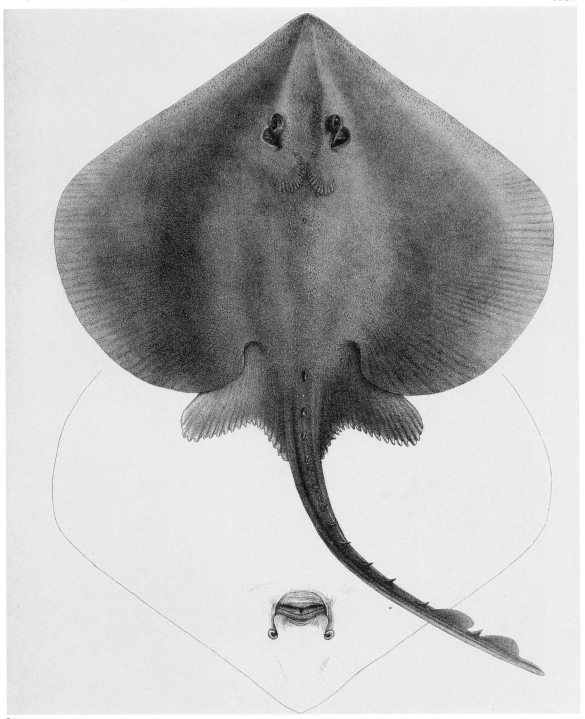

RAJA MICROPS.
(Mouth of the Rio de la Plata)

A. MICROPOGON ORNATUS. B. LÆMONEMA LONGIFILIS.
(Mouth of the Rio de la Plata.)

The Voyage of "H.M.S. Challenger." Shore Fishes Pl. VI.

R. Mintern del et lith.

RAJA BRACHYURA.
(Magelhæn's Straits)

Mintern Bros imp.

Shore Fishes. Pl. V.

RAYI.

Mintern Bros imp.

The Voyage of H.M.S. "Challenger".

R. Mintern del.

RAJA MU
(Kerguelen)

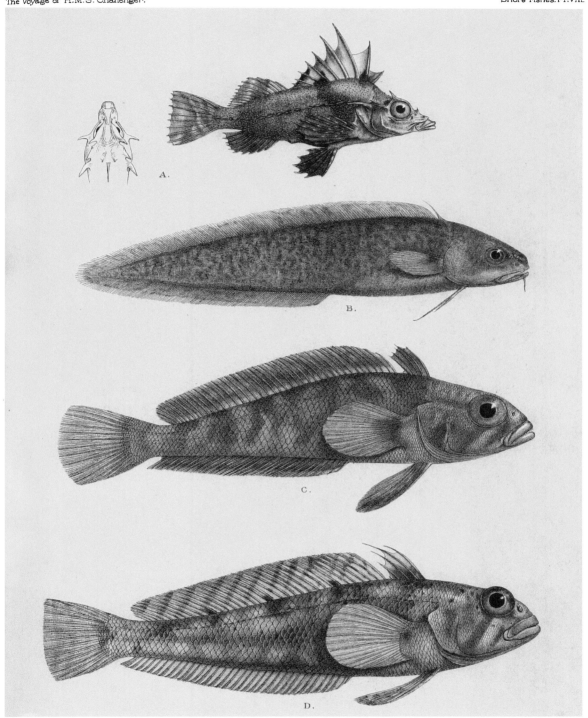

A. ZANCLORHYNCHUS SPINIFER. B. MURÆNOLEPIS MARMORATA.
C. NOTOTHENIA SQUAMIFRONS D. NOTOTHENIA MIZOPS.
(Kerguelen's Land.)

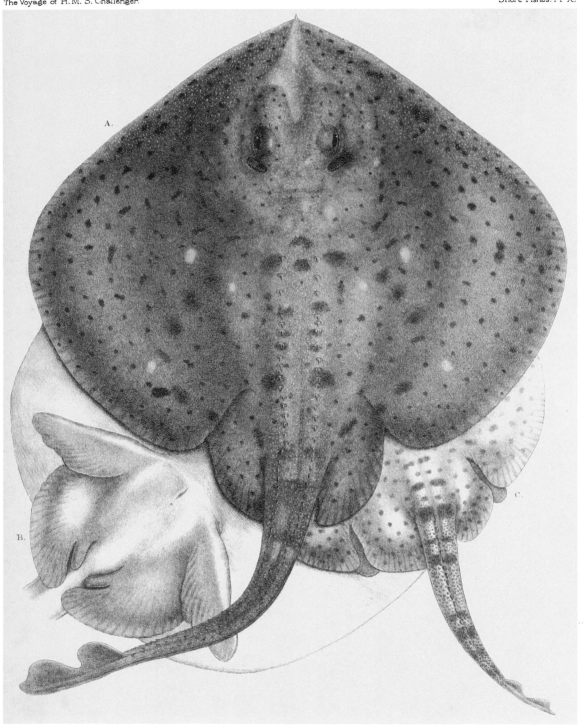

PSAMMOBATIS RUDIS.
(Magelhæns Straits.)

A. THYSANOPSETTA NARESII. B. LYCODES MACROPS. C. NOTOTHENIA ELEGANS.
(Magelhan's Straits.)

SCORPÆNA THOMSONII.
(Juan Fernandez.)

A. TRIGLA PICTA. B. UMBRINA REEDII.
(Juan Fernandez.)

A. RAJA NITIDA. B. SOLENOGNATHUS FASCIATUS.
(Twofold Bay.)

A. LÆOPS PARVICEPS. B. LOPHONECTES GALLUS. C. CALLIONYMUS PHASIS.
(Twofold Bay.)

A. APOGON SEPTEMSTRIATUS. B. APOGON MONOGRAMMA. C. APOGON ARAFURÆ.
D. CENTROPRISTIS PLEUROSPILUS. E. ANTHIAS MEGALEPIS.
(Arafura Sea.)

A. PLATYCEPHALUS SCULPTUS. B. SEBASTES HEXANEMA. C. LIOSCORPIUS LONGICEPS.
(Arafura Sea.)

A. ACANTHAPHRITIS GRANDISQUAMIS. B. TRIGLA LEPTACANTHUS.
C. LEPIDOTRIGLA SPILOPTERA. D. MINOUS PICTUS.
(Arafura Sea.)

A. URANOSCOPUS KAIANUS. B. CALLIONYMUS KAIANUS.
(Kay Islands.)
C. TETRABRACHIUM OCELLATUM.
(South Coast of New Guinea.)

A. OPHIDIUM MURÆNOLEPIS. B. PROPOMA ROSEUM. C. XIPHOCHILUS QUADRIMACULATUS.
D. HELIASTES ROSEUS.
(Kai Islands.)

A. RHOMBOIDICHTHYS SPILURUS. B. RHOMBOIDICHTHYS ANGUSTIFRONS.
C. SOLEA KAIANA. D. SAMARIS MACULATUS.
(Arafura Sea.)

A. ANTICITHARUS POLYSPILUS. B. POECILOPSETTA COLORATA.
(Kai Islands.)

A. CHAMPSODON VORAX. B. MONACANTHUS TESSELLATUS.
(Philippine Islands.)
C. SAURUS KAIANUS. D. MONACANTHUS FILICAUDA.
Arafura Sea.

A.B. PSEUDORHOMBUS OCELLATUS ♂ et ♀. C. NEMATOPS MICROSTOMA.
(Admiralty Islands.)

A.B. JULIS OBSCURA ad. & juv. C. SICYDIUM NIGRESCENS. D. DORYICHTHYS PLEUROTÆNIA.
(Sandwich Islands.)

SEBASTES MACROCHIR.
(Japan.)

SEBASTES OBLONGUS.
(Japan)

A. SEBASTES JOYNERI. B. PLATYCEPHALUS RUDIS.
(Japan.)

A. CYNOGLOSSUS JOYNERI.   B. CYNOGLOSSUS INTERRUPTUS.
C. LEPIDOPSETTA MACULATA. *(Japan.)*   D. HELIASTES FLAVICAUDA. *(Prince Edward Id.)*
(Pernambuco.)

A. SALMO MACROSTOMA. B. LEUCISCUS HAKUENSIS.
(Japan.)

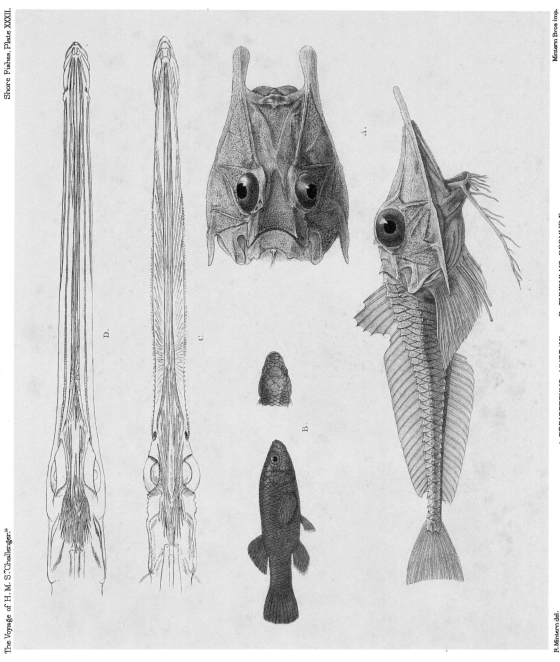

A. PERISTETHUS MURRAYI. B. FUNDULUS BERMUDÆ. (*Bermuda.*)
C. FISTULARIA SERRATA. D. FISTULARIA DEPRESSA.

第二卷 动物志二彩图

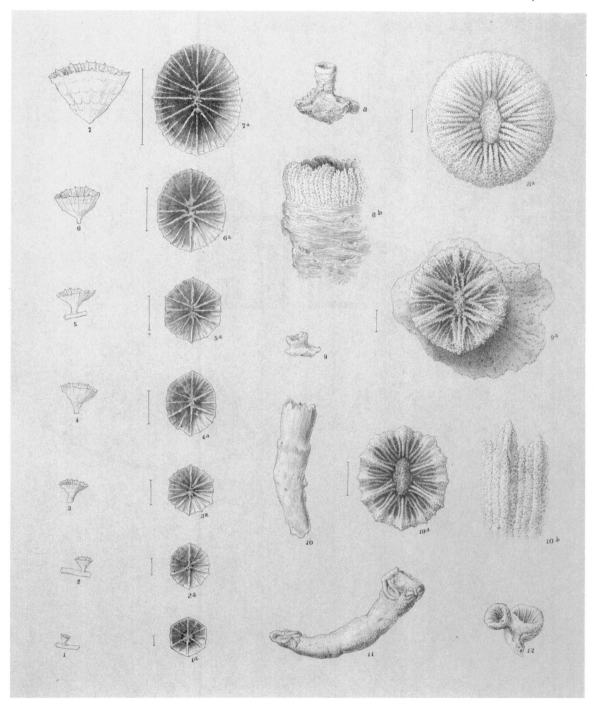

1-7 FLABELLUM. 8 THECOPSAMMIA. 9-12 BALANOPHYLLIA

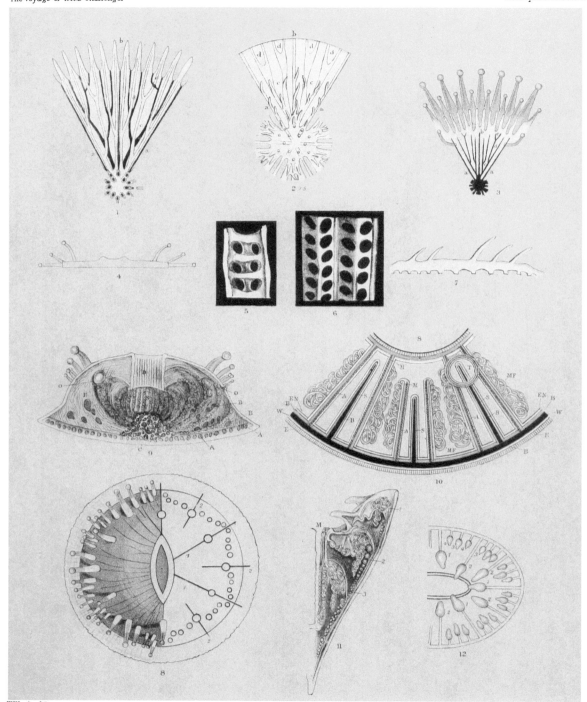

LEPTOPENUS, STEPHANOPHYLLIA, FLABELLUM.

LORICULUS PANAYENSIS, 1 ♂, 2 ♀.

BATRACHOSTOMUS SEPTIMUS.

BUCEROS MINDANENSIS.

DICRURUS STRIATUS

1. DICÆUM MINDANENSE.   2,3. NECTAROPHILA JULIÆ, ♂ & ♀.

PHABOTRERON BREVIROSTRIS.

1. MONARCHA INFELIX.   2. RHIPIDURA SEMIRUBRA.
3. MYZOMELA PAMMELÆNA.

CARPOPHAGA RHODINOLÆMA.

PTILOPUS JOHANNIS ♂ et ♀.

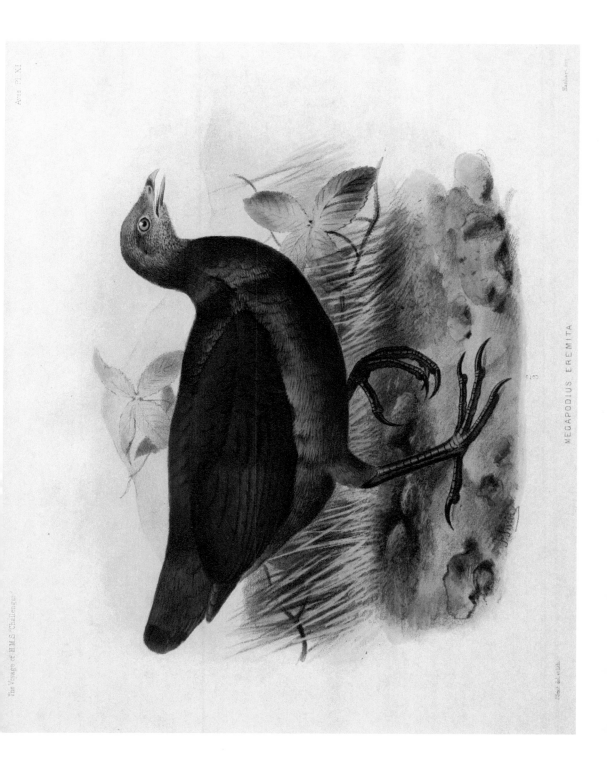

MEGAPODIUS EREMITA

1. PTILOTIS CARUNCULATA
2. " PROCERIOR

1. PTILOTIS PROVOCATOR.
2. " " jr.

1. ZOSTEROPS FLAVICEPS.
2. ,, EXPLORATOR.

CHRYSŒNAS VICTOR. ♂

CHRYSŒNAS VIRIDIS.

CARPOPHAGA LATRANS.

1. PACHYCEPHALA PHÆONOTA. 2. RHIPIDURA SQUAMATA.

GRAUCALUS POLLENS.

TRICHOGLOSSUS NIGRIGULARIS.

BUTEO SOLITARIUS.

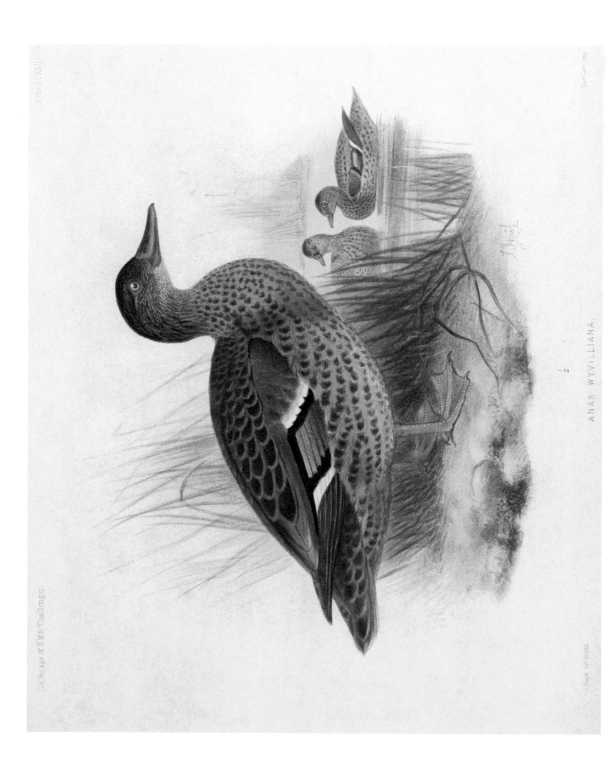
ANAS WYVILLIANA

NESOCICHLA EREMITA.

NESOSPIZA ACUNHÆ.

1. PHALACROCORAX IMPERIALIS.
2. ALBIVENTRIS.

1. PHALACROCORAX VERRUCOSUS, ♂.
2. "   "   , jr.

SPHENISCUS DEMERSUS.

1. SPHENISCUS MAGELLANICUS.
2. " " jr.

1. EUDYPTES CHRYSOLOPHUS.
2.   "   "   , pale variety.

1. EUDYPTES CHRYSOCOME.
2. " " .pull.

第三卷动物志三彩图

The Voyage of H.M.S. "Challenger". Pycnogonida. Pl. XVI.

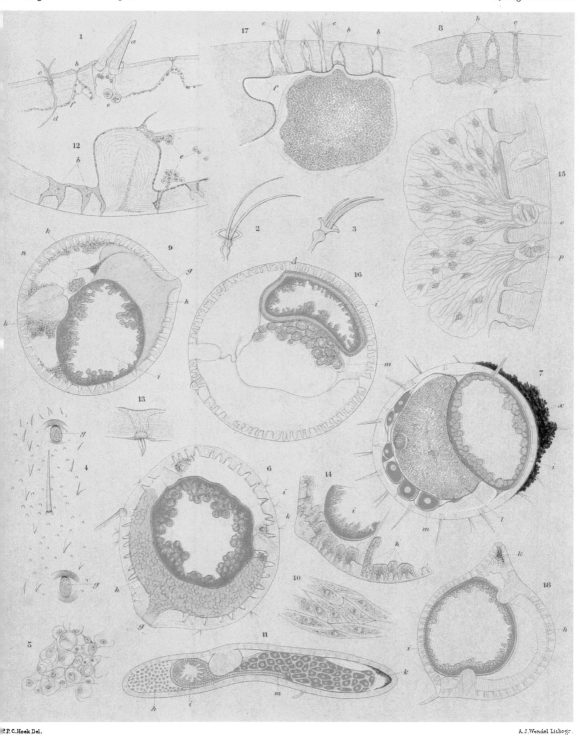

1-7 NYMPHON. 8-11 ASCORHYNCHUS. 12-16 COLOSSENDEIS. 17-18 PHOXICHILIDIUM.

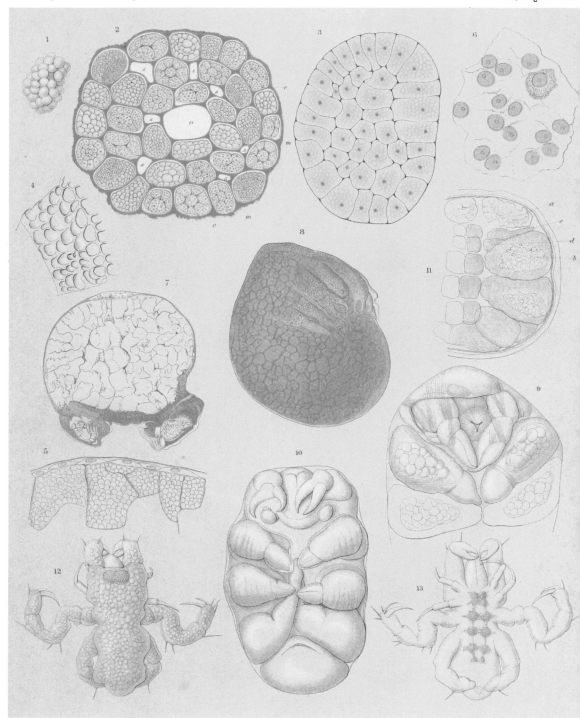

EMBRYOLOGY OF NYMPHON.

第四卷 动物志四彩图

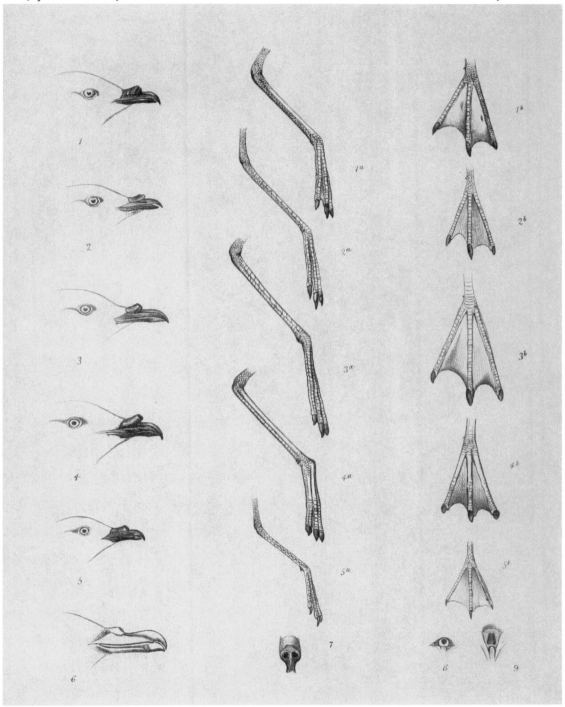

EXTERNAL CHARACTERS OF PETRELS.
(Oceanitidæ, &c.)

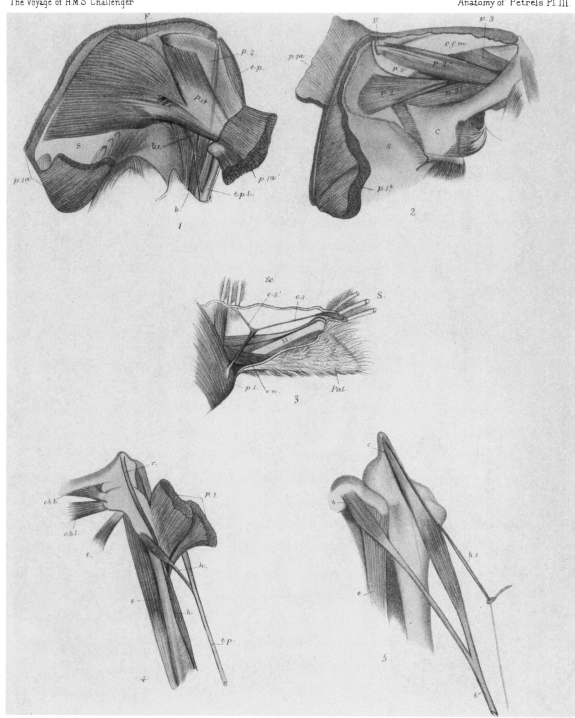

MYOLOGY OF PETRELS.
Anterior Extremity

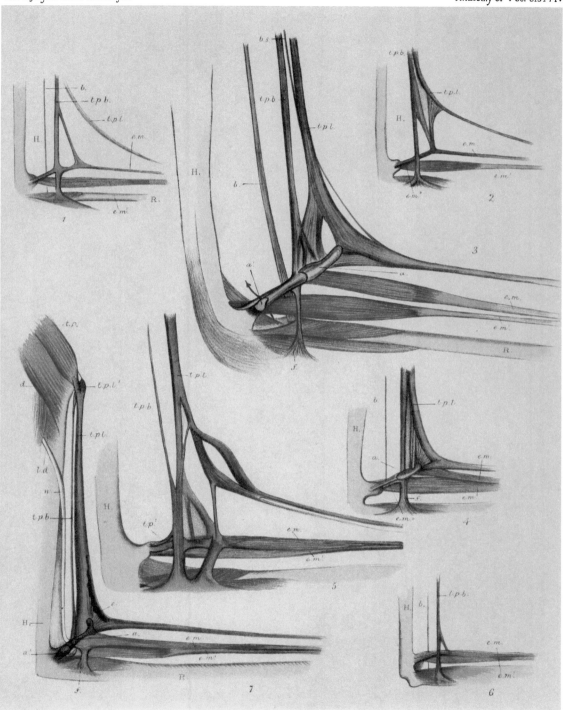

MYOLOGY OF PETRELS.
Tensor Patagii muscles.

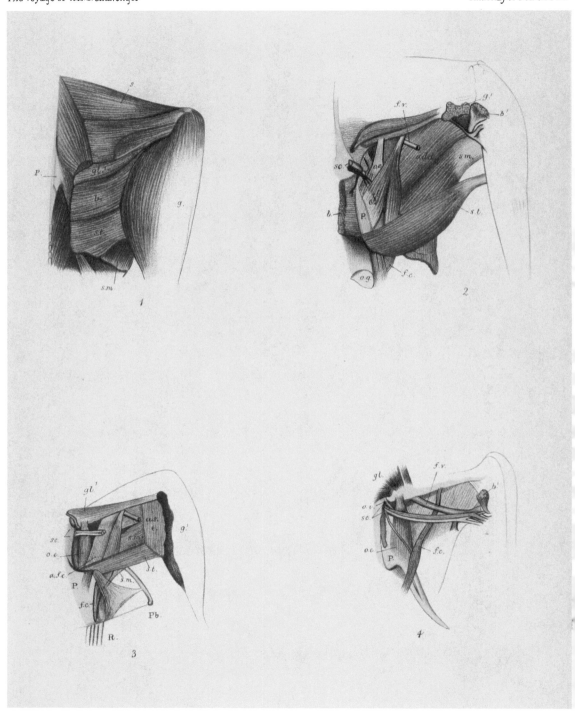

MYOLOGY OF PETRELS.
Hinder Extremity.

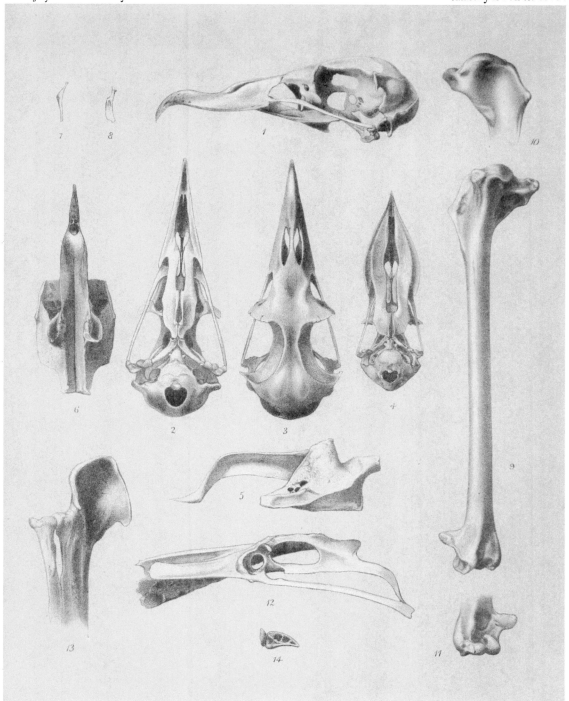

OSTEOLOGY OF PETRELS
Skull, Pelvis and Limb-bones.

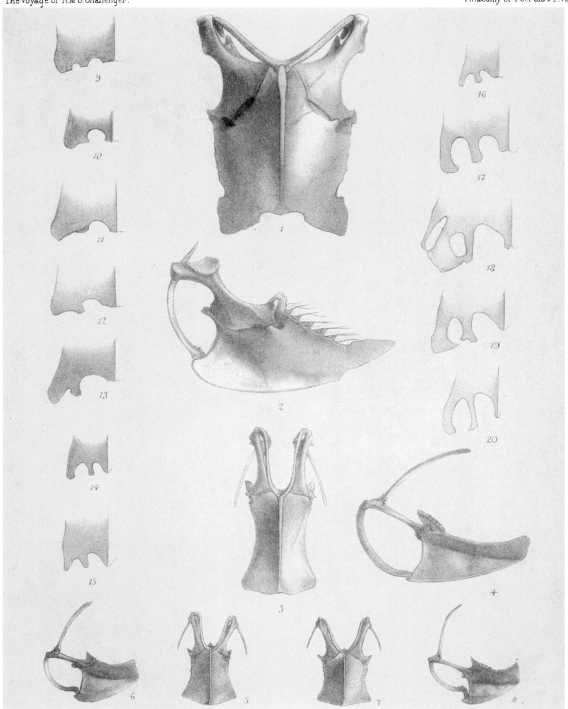

OSTEOLOGY OF PETRELS.
Sterna.

THAMNOSTYLUS DINEMA.

The Voyage of H.M.S. "Challenger". Deep-Sea-Medusae. Pl. 2.

E. Haeckel and A. Giltsch, Del. E. Giltsch, Jena, Lith. gr.

PTYCHOGENA PINNULATA.

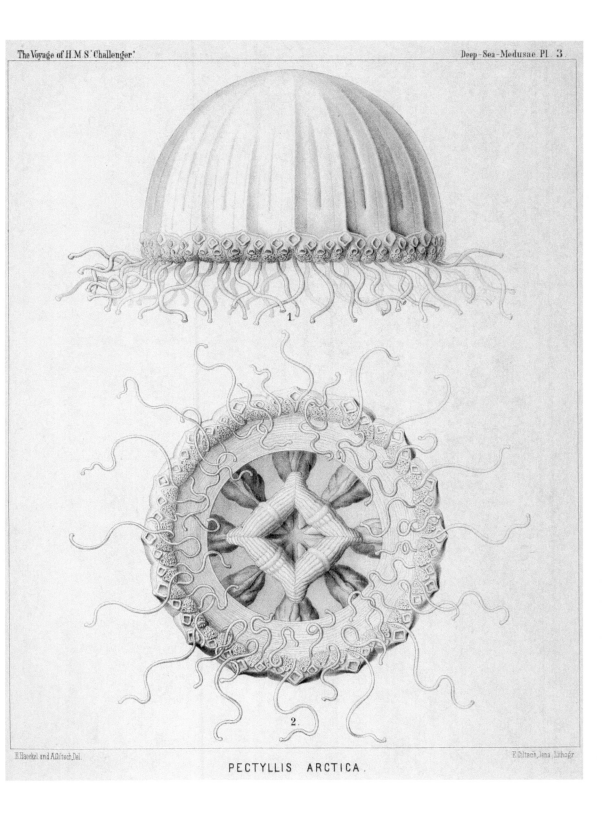
PECTYLLIS ARCTICA.

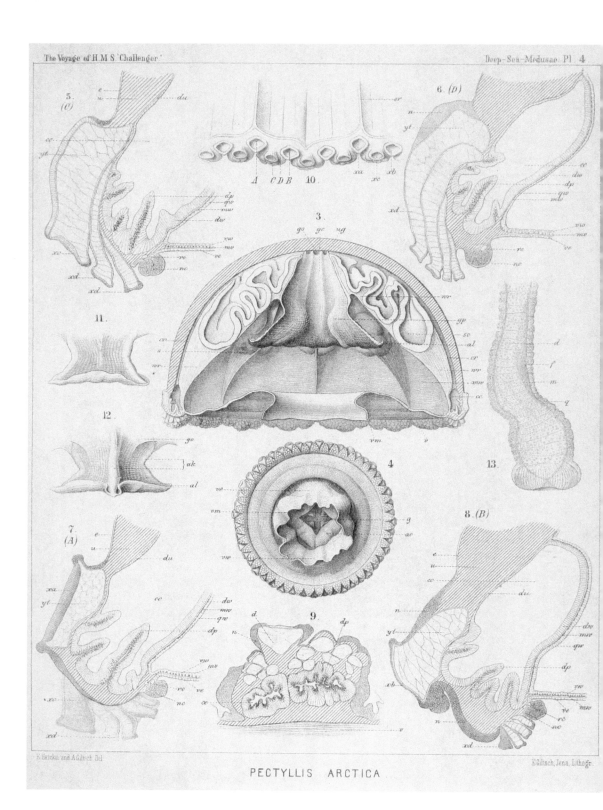

PECTYLLIS ARCTICA

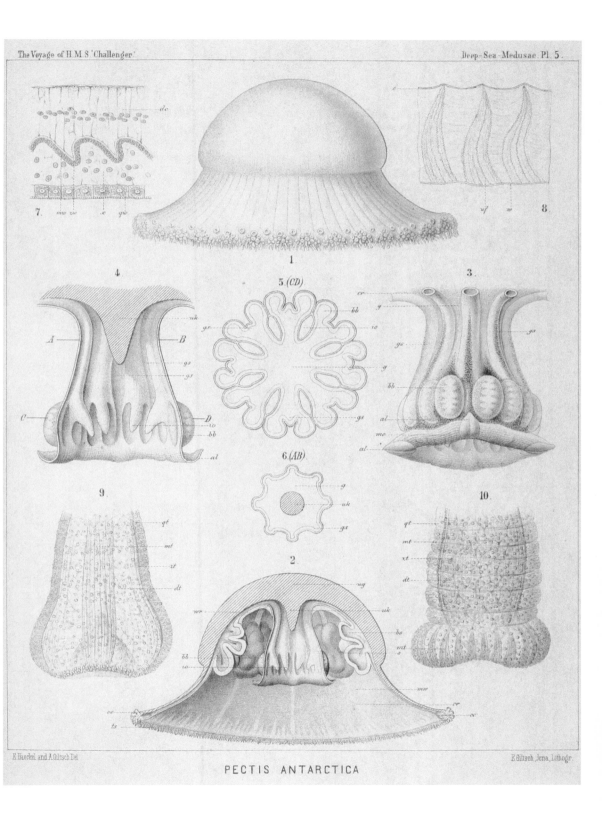

PECTIS ANTARCTICA

PECTANTHIS ASTEROIDES.

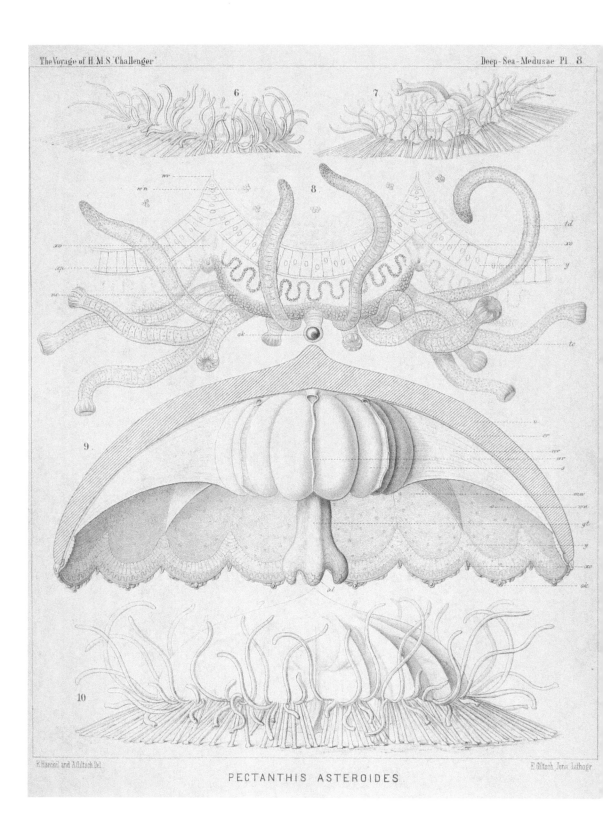

PECTANTHIS ASTEROIDES.

CUNARCHA AEGINOIDES.

POLYCOLPA FORSKALII.

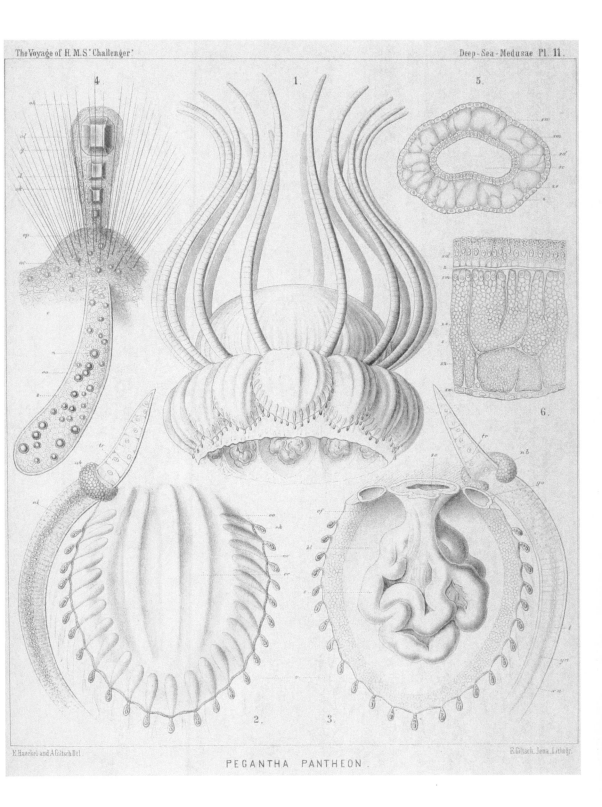

PEGANTHA PANTHEON.

PEGANTHA PANTHEON.

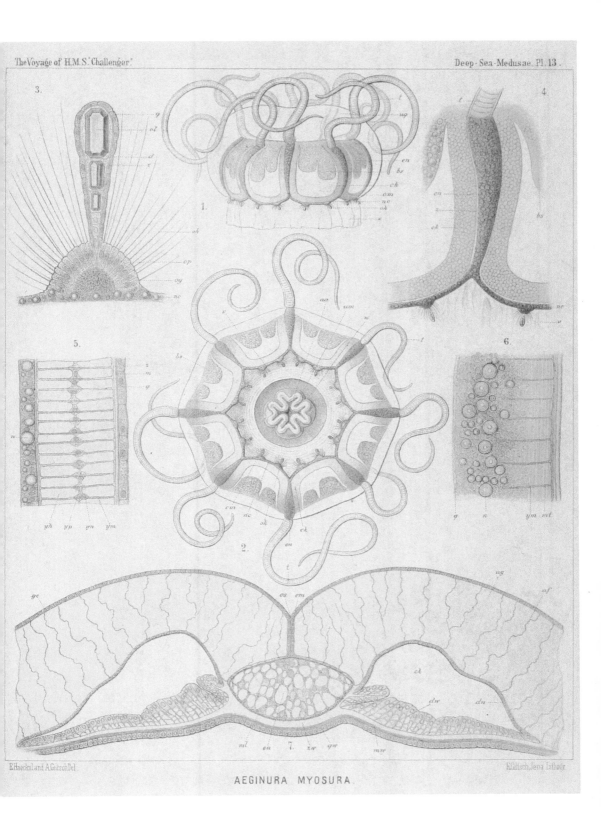

AEGINURA MYOSURA.

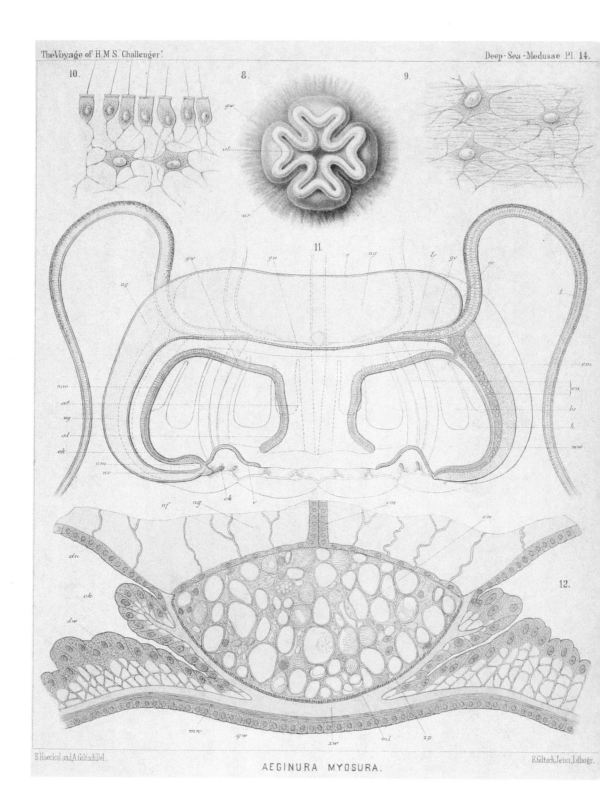

AEGINURA MYOSURA.

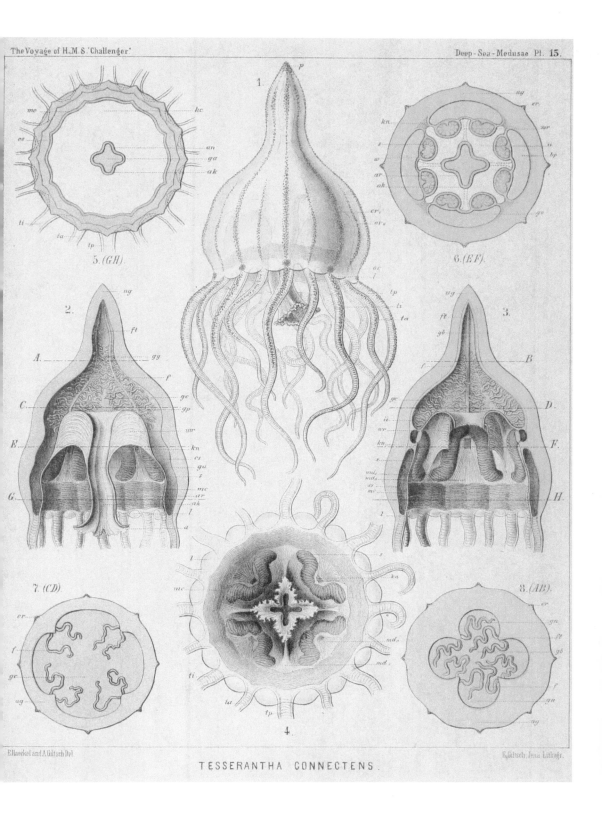

TESSERANTHA CONNECTENS.

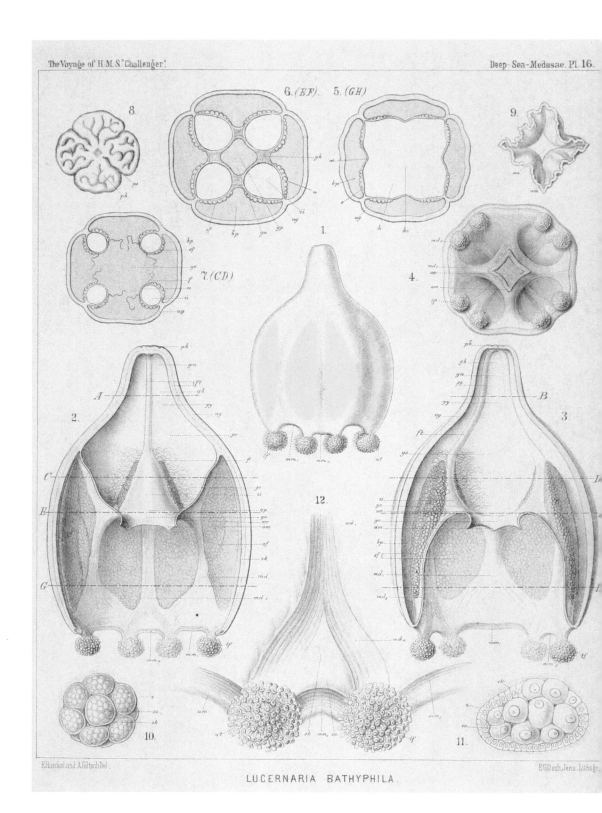

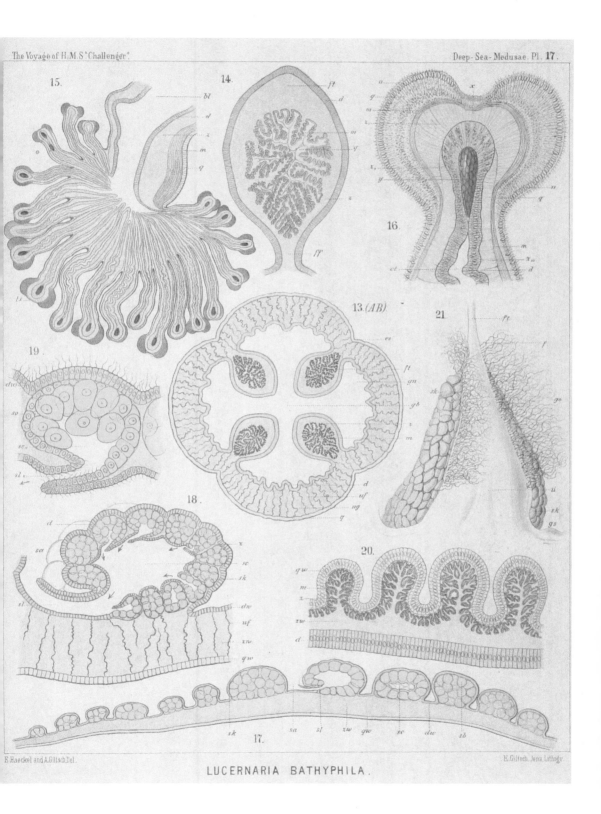

LUCERNARIA BATHYPHILA.

PERIPHYLLA MIRABILIS.

PERIPHYLLA MIRABILIS.

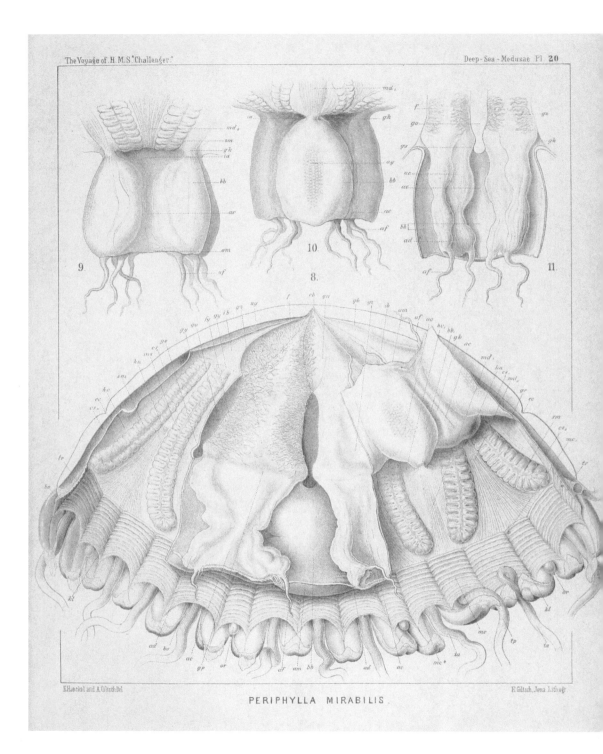

PERIPHYLLA MIRABILIS.

PERIPHYLLA MIRABILIS.

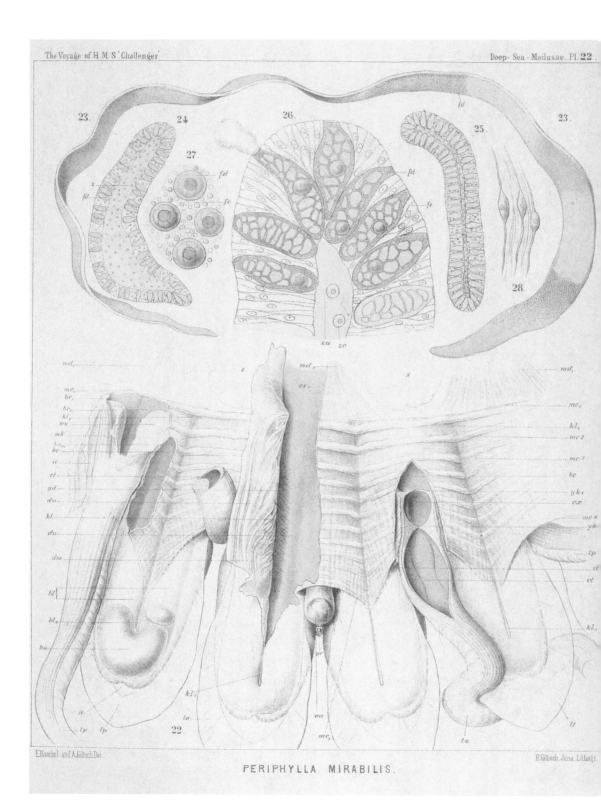

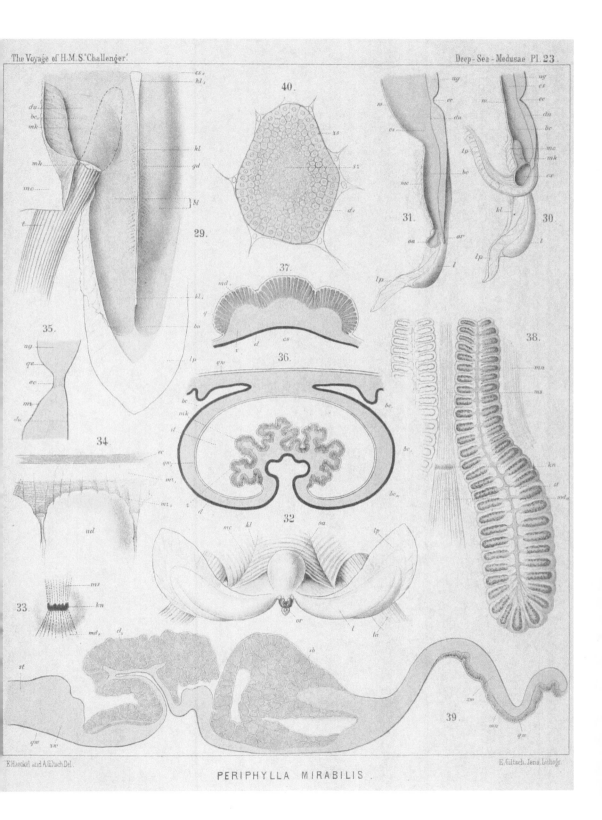

PERIPHYLLA MIRABILIS.

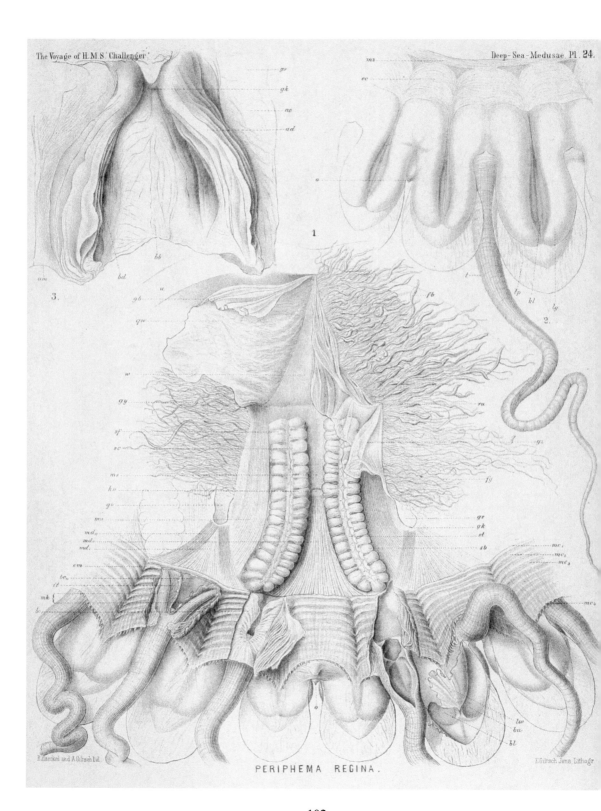
PERIPHEMA REGINA.

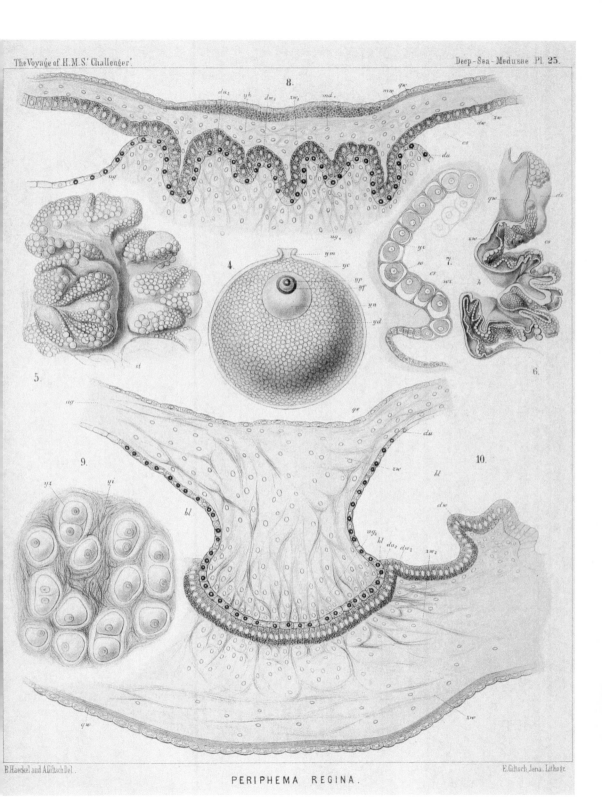

PERIPHEMA REGINA.

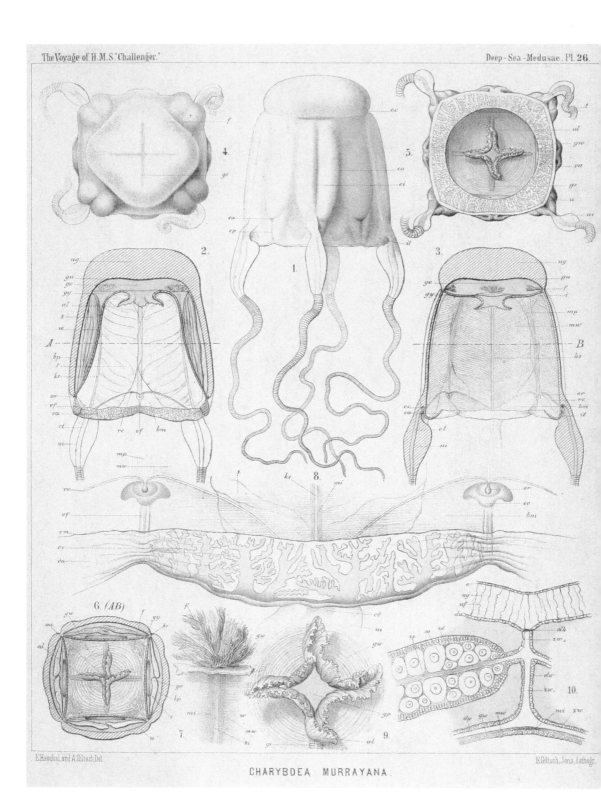

CHARYBDEA MURRAYANA.

NAUPHANTA CHALLENGERI.

NAUPHANTA CHALLENGERI.

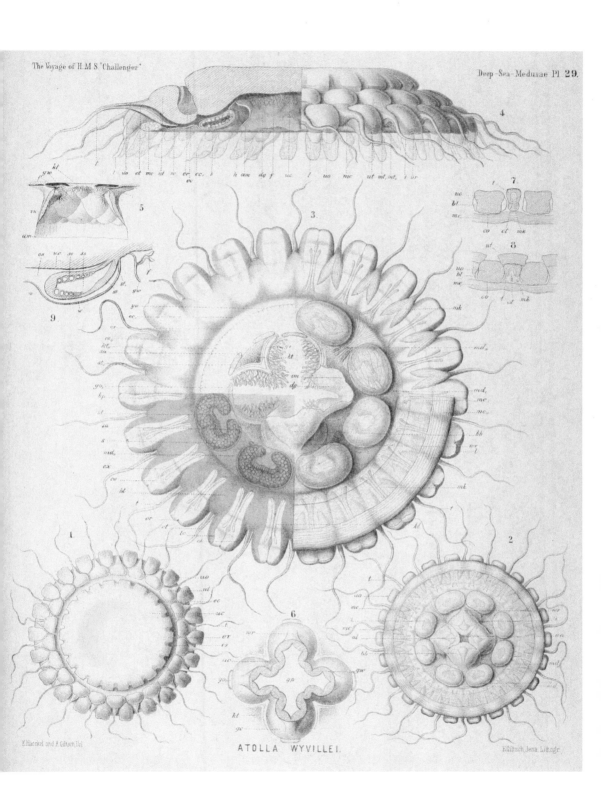

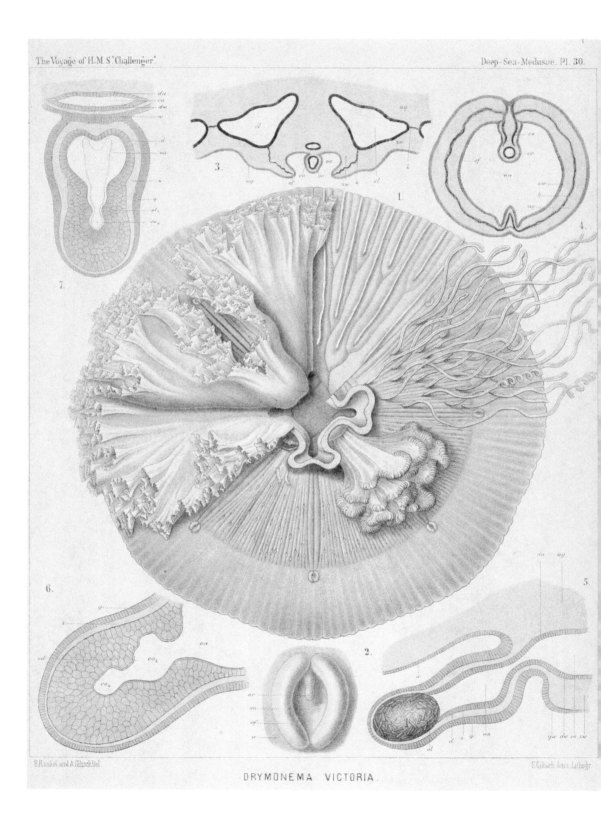

DRYMONEMA VICTORIA

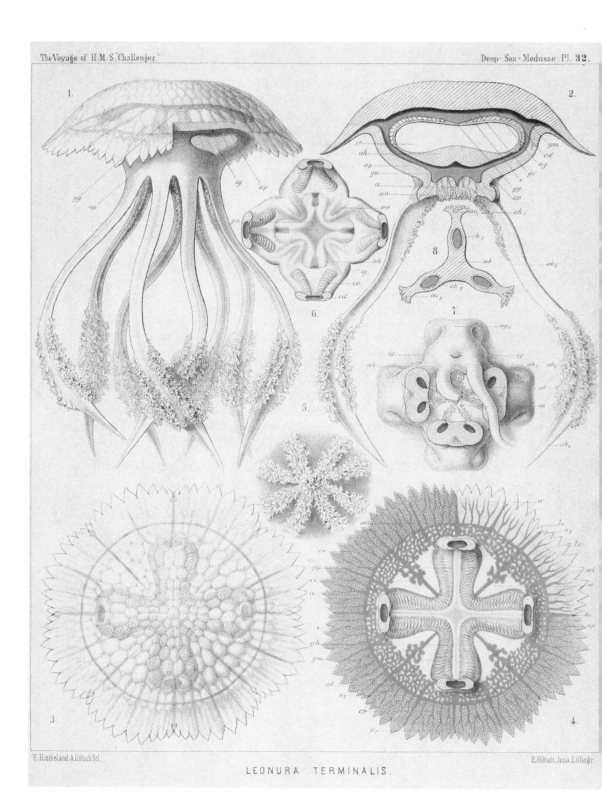

LEONURA TERMINALIS.

第六卷动物志六彩图

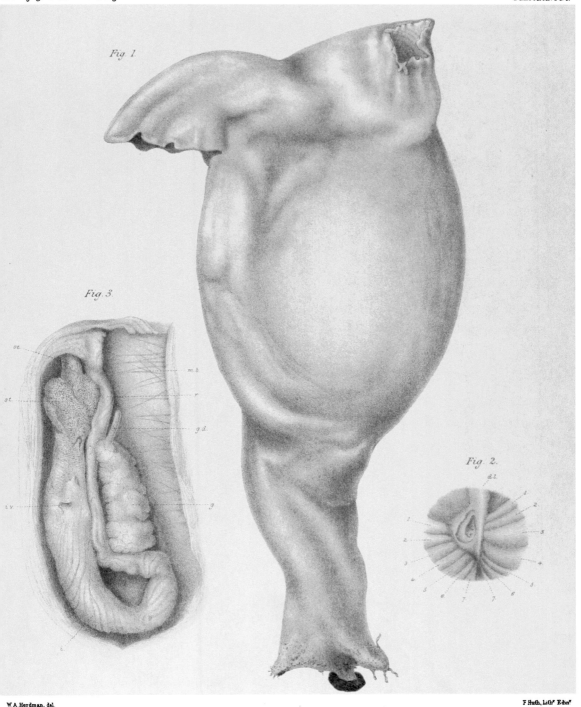

ASCOPERA GIGANTEA, Herdman.

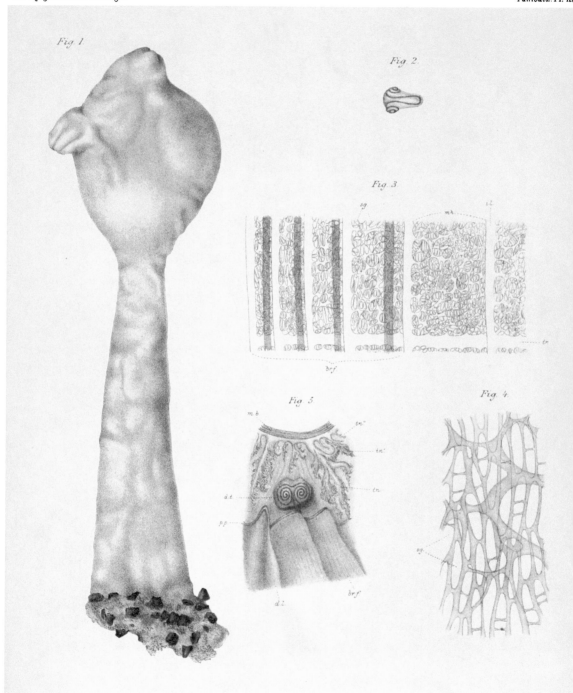

FIGS. 1-2 ASCOPERA PEDUNCULATA, Herdman.   FIGS. 3-5 ASCOPERA GIGANTEA, Herdman.

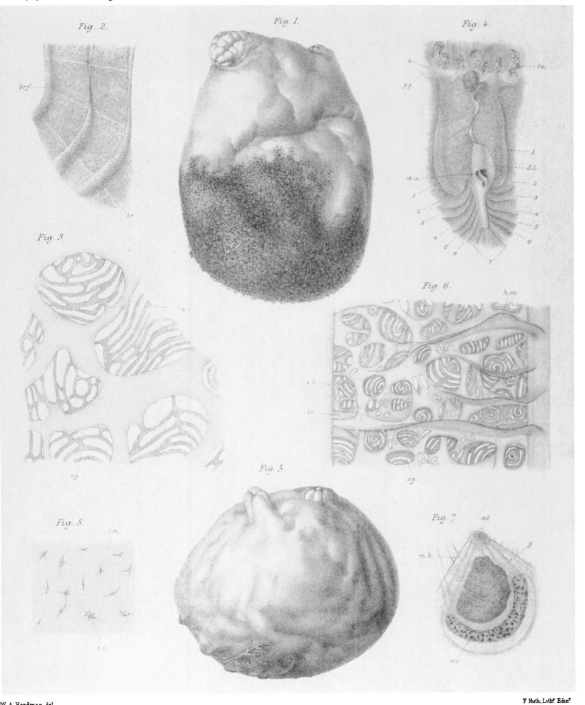

FIGS. 1-4 MOLGULA GIGANTEA, Cunningham.   FIGS. 5-8 MOLGULA GREGARIA, Lesson.

The Voyage of H.M.S."Challenger". Tunicata Pl. V

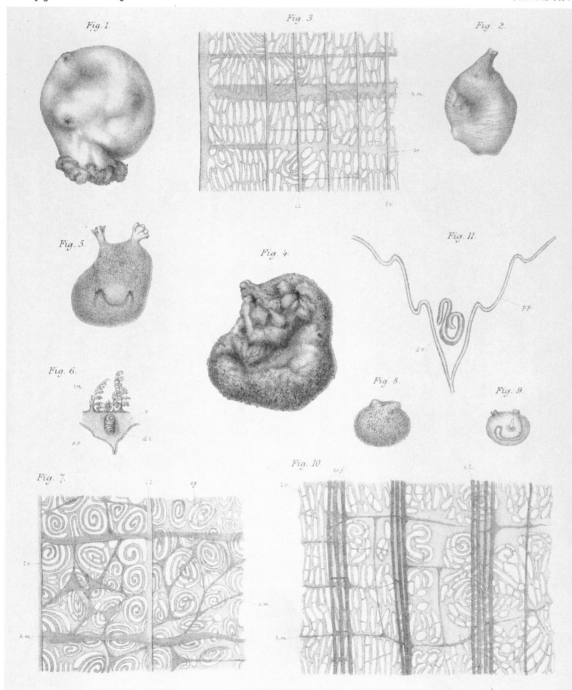

W. A. Herdman, del. F. Huth, Lith'. Edin'.

FIGS 1-3 MOLGULA PEDUNCULATA, Herdman. FIGS 4-7 MOLGULA HORRIDA, Herdman.
FIGS 8-11 MOLGULA FORBESI, Herdman.

116

第七卷动物志七彩图

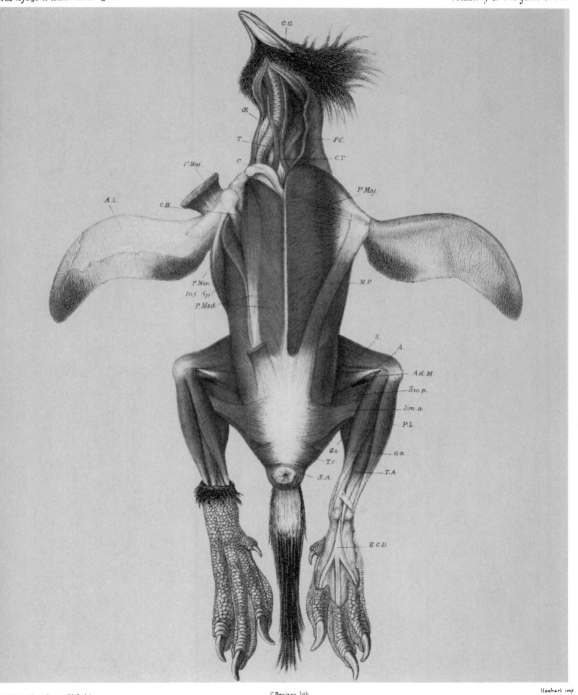

MUSCLES OF EUDYPTES CHRYSOCOME.

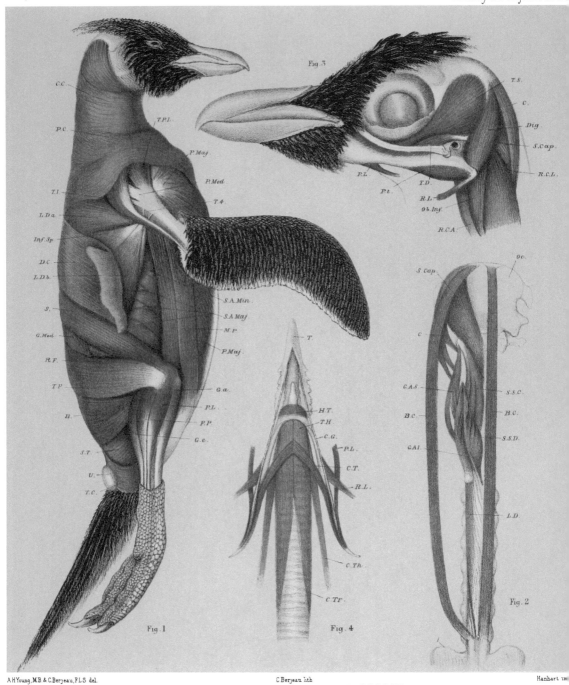

MUSCLES OF EUDYPTES CHRYSOCOME.

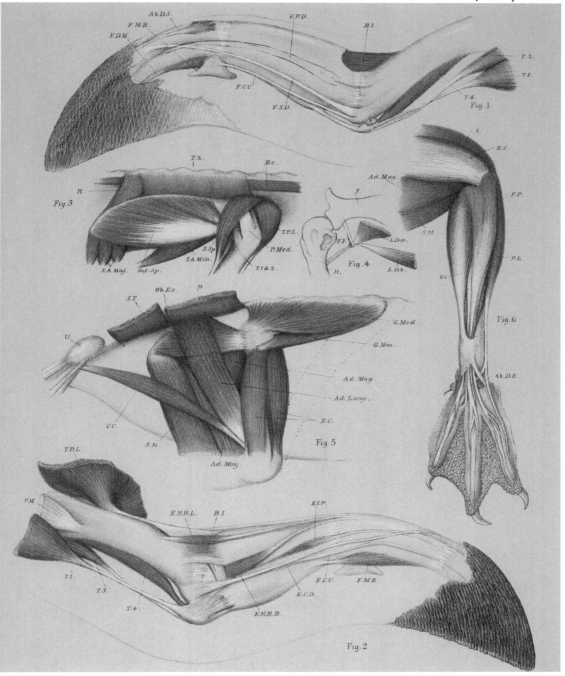

MUSCLES OF EUDYPTES CHRYSOCOME.

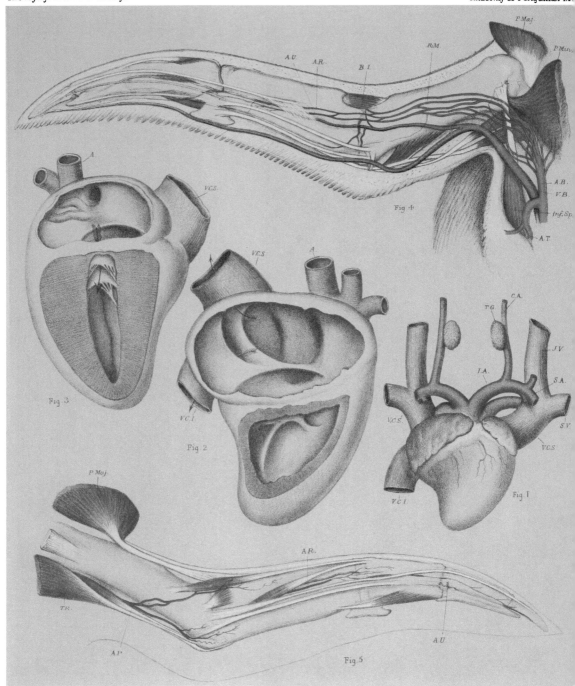

ORGANS OF CIRCULATION.

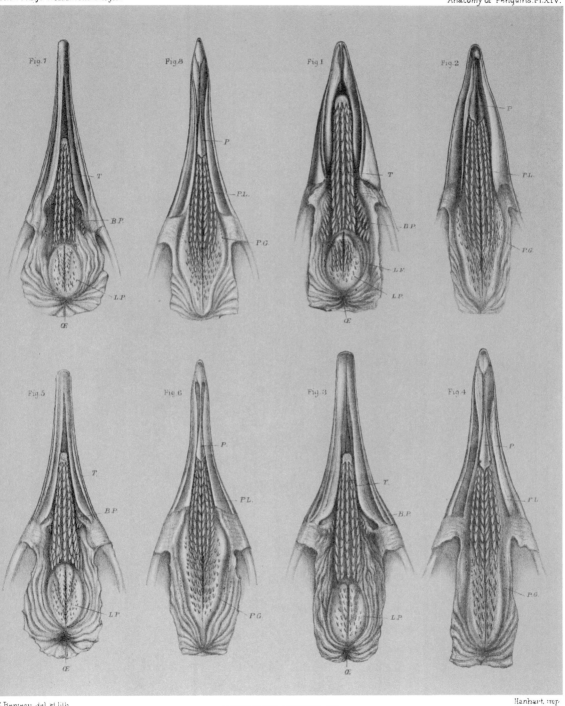

DIGESTIVE ORGANS OF
1,2, EUDYPTES CHRYSOLOPHUS.   3,4, SPHENISCUS MAGELLANICUS.
5,6, S. DEMERSUS.   7,8, S. MENDICULUS.

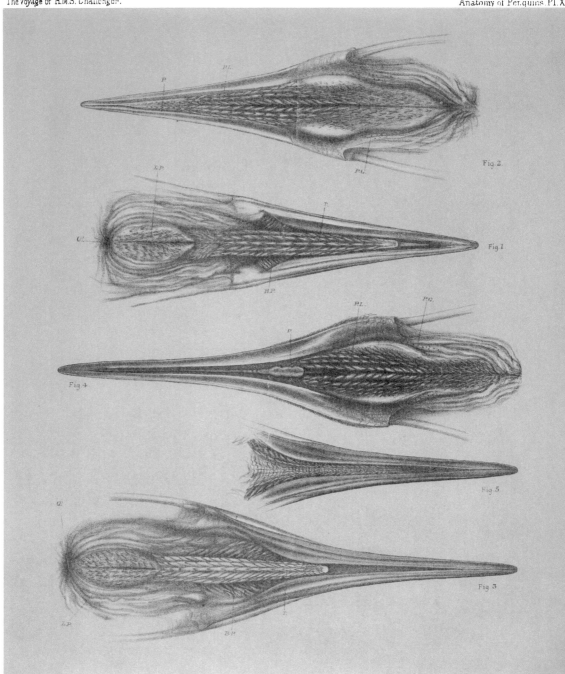

DIGESTIVE ORGANS OF
1.2. PYGOSCELES TÆNIATUS. 3.4.5. APTENODYTES LONGIROSTRIS.

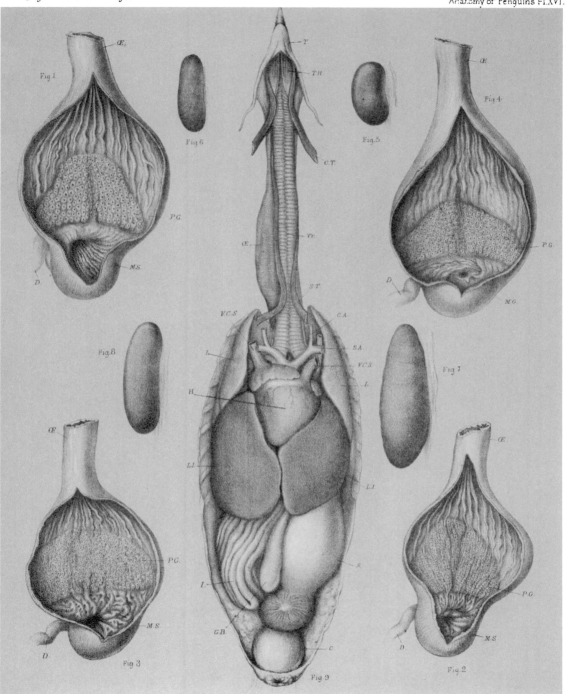

DIGESTIVE ORGANS OF
5, EUDYPTES CHRYSOCOME 6,1, E.CHRYSOLOPHUS. 2,7,9, SPHENISCUS DEMERSUS.
3,8, S.MAGELLANICUS. 4, S.MENDICULUS.

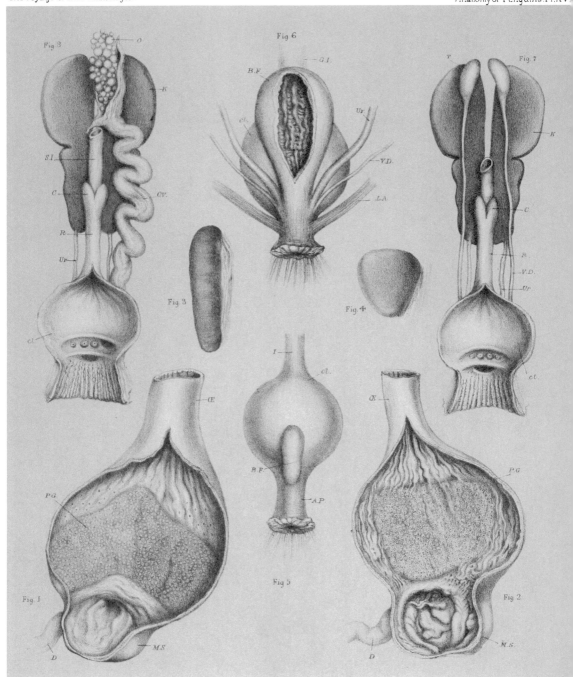

VISCERA OF
1,3, PYGOSCELES TÆNIATUS. 2,4,6, APTENODYTES LONGIROSTRIS.
5,7,8, EUDYPTES CHRYSOCOME.

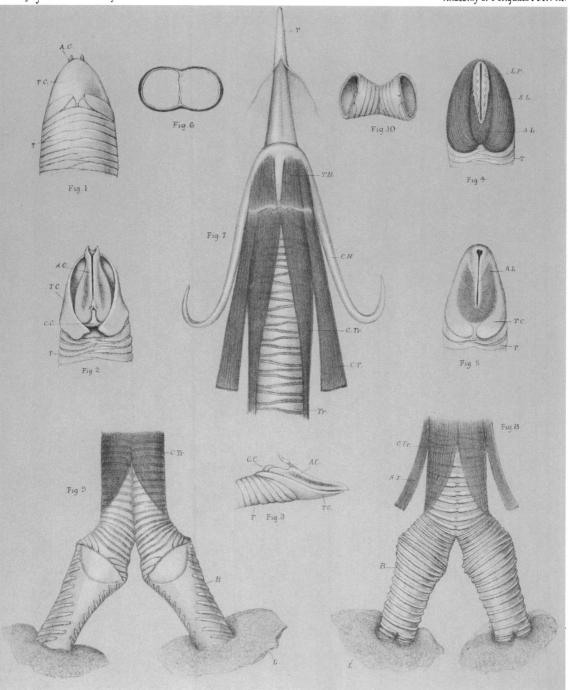

RESPIRATORY ORGANS OF APTENODYTES LONGIROSTRIS.

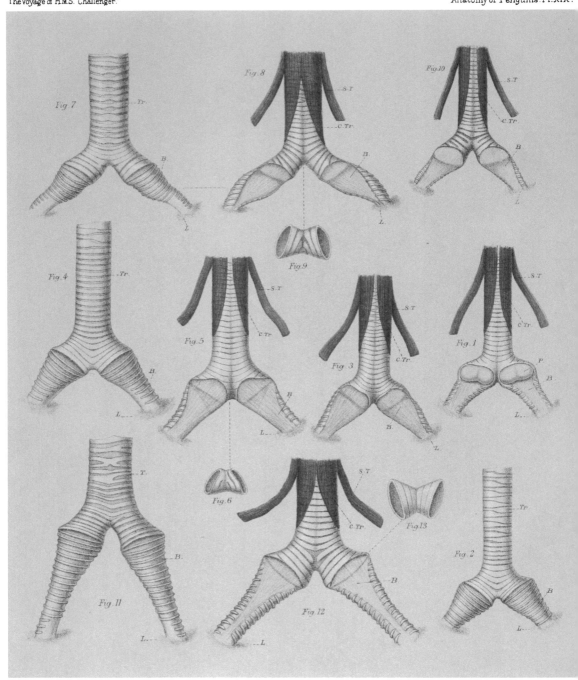

RESPIRATORY ORGANS OF
1.2. EUDYPTES CHRYSOCOME, From Tristan. 3. E. CHRYSOCOME, From Falklands.
4.5.6. E. CHRYSOLOPHUS. 7.8.9. SPHENISCUS MAGELLANICUS.
10. S. MENDICULUS. 11.12.13. PYGOSCELES TÆNIATUS.

LYTOCARPUS SECUNDUS.

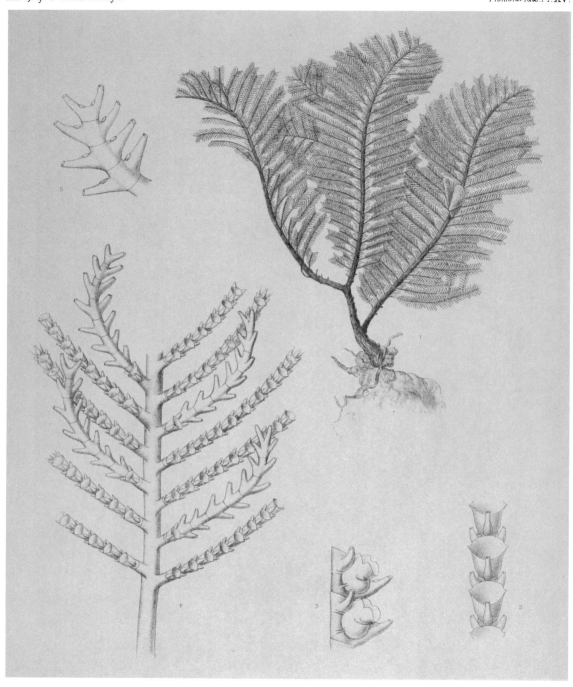

LYTOCARPUS SPECTABILIS.

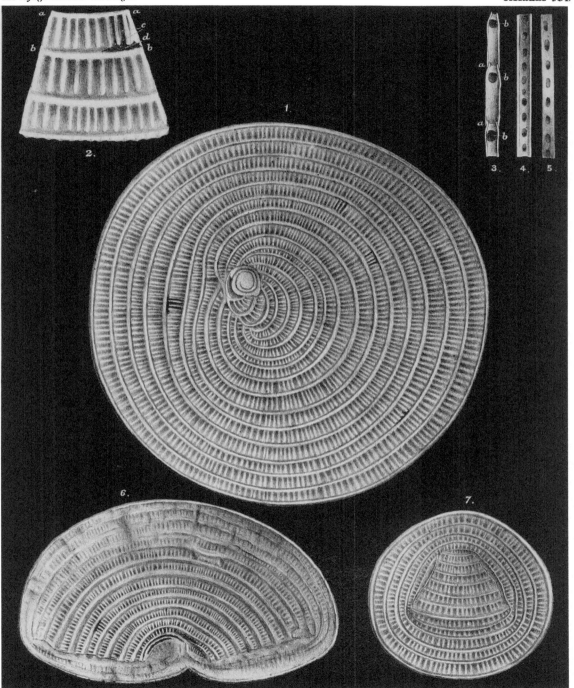

ORBITOLITES TENUISSIMA.

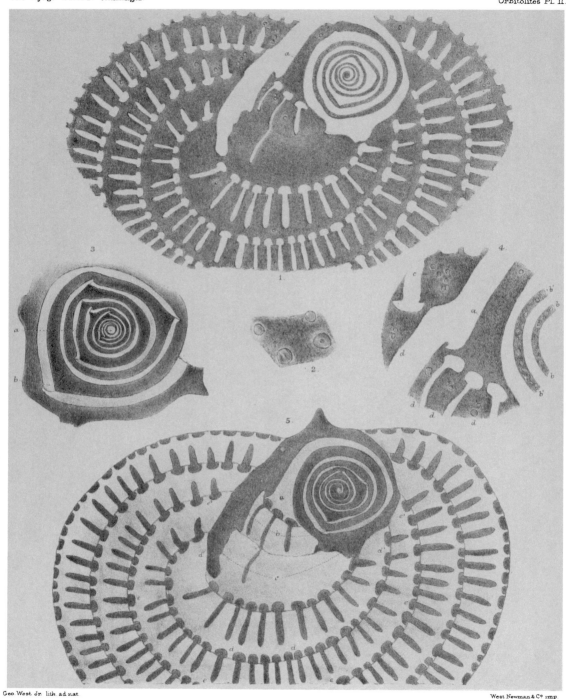

ORBITOLITES TENUISSIMA.

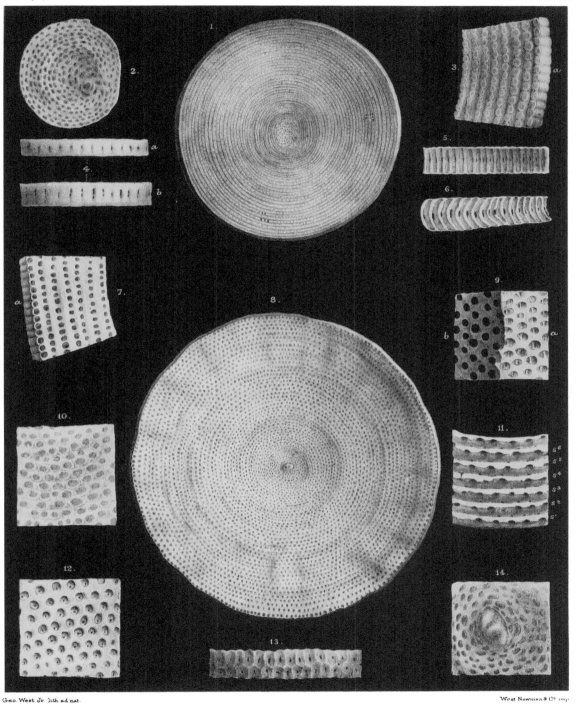

FIGS. 1–7. ORBITOLITES MARGINALIS, FIGS. 8–14. O. DUPLEX.

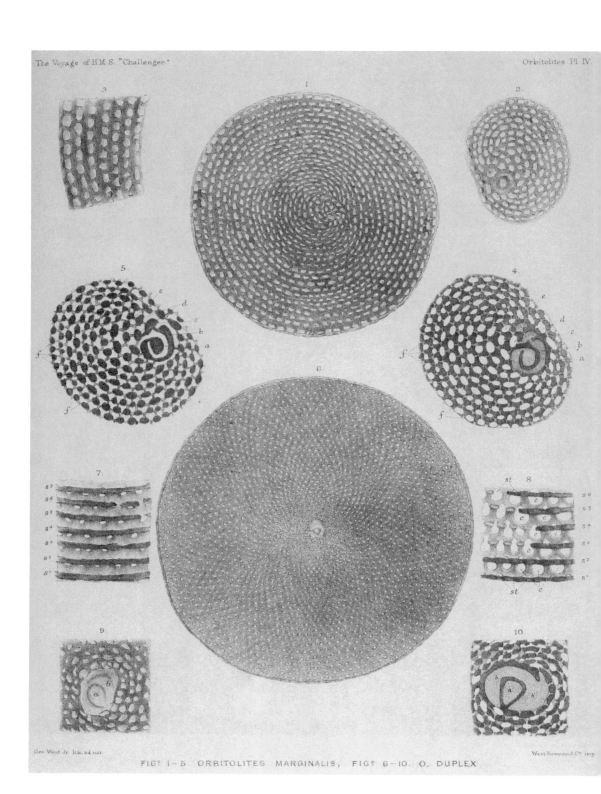

FIGS. 1–5. ORBITOLITES MARGINALIS, FIGS. 6–10. O. DUPLEX

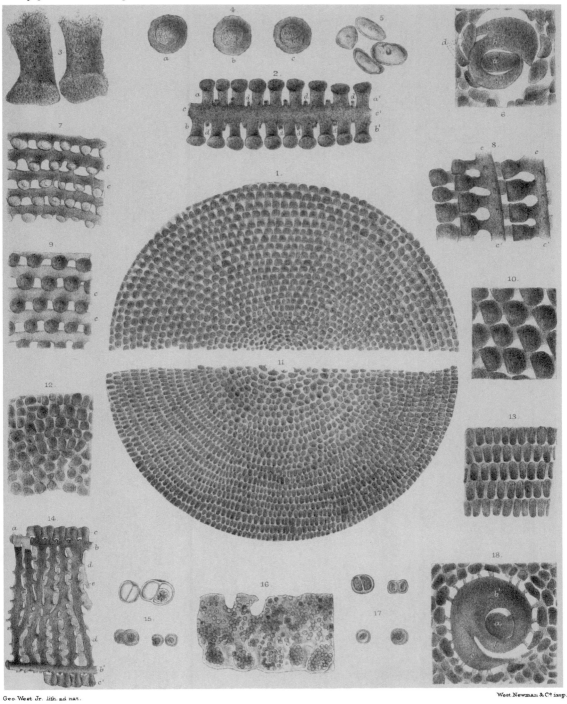

FIGS 1—10. ORBITOLITES DUPLEX. FIGS 11—18. O. COMPLANATA.

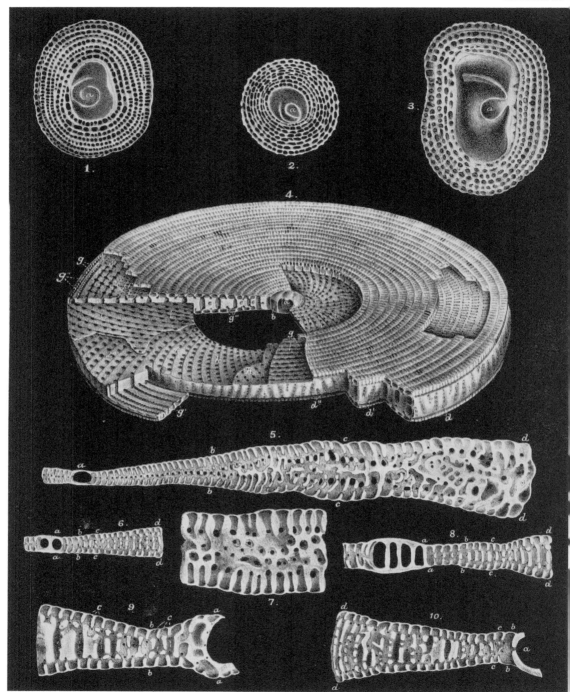

ORBITOLITES COMPLANATA.

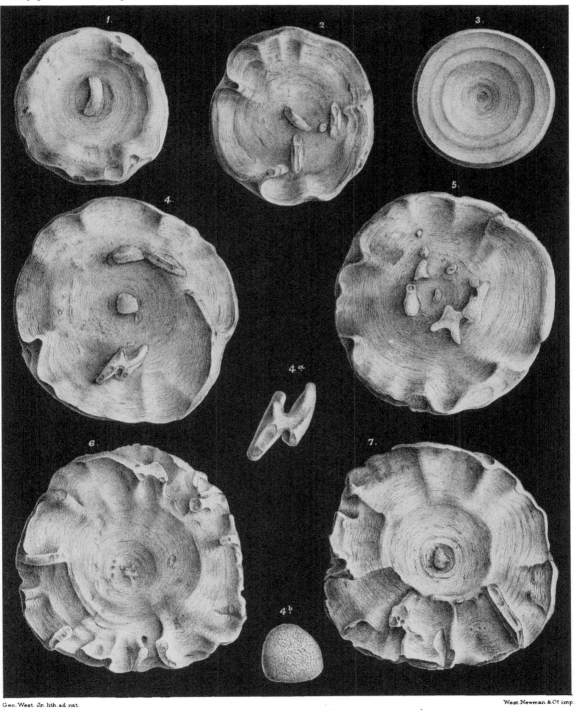

ORBITOLITES COMPLANATA. (VAR. LACINIATA.)

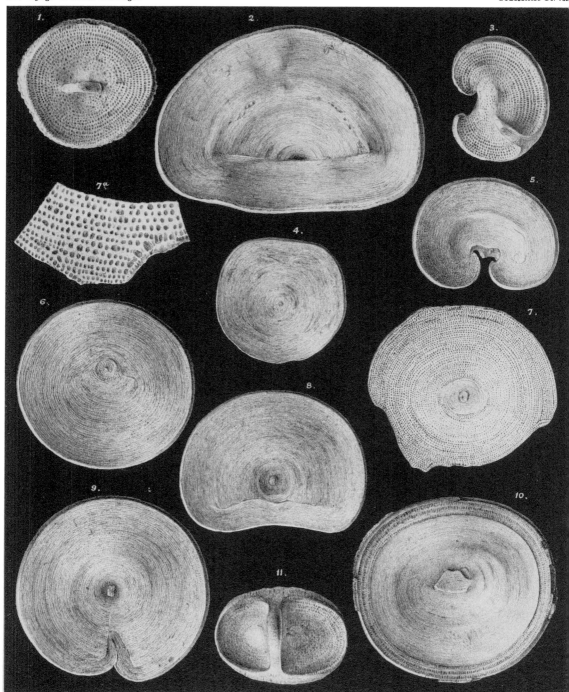

ORBITOLITES COMPLANATA. — IRREGULARITIES.

第八卷动物志八彩图

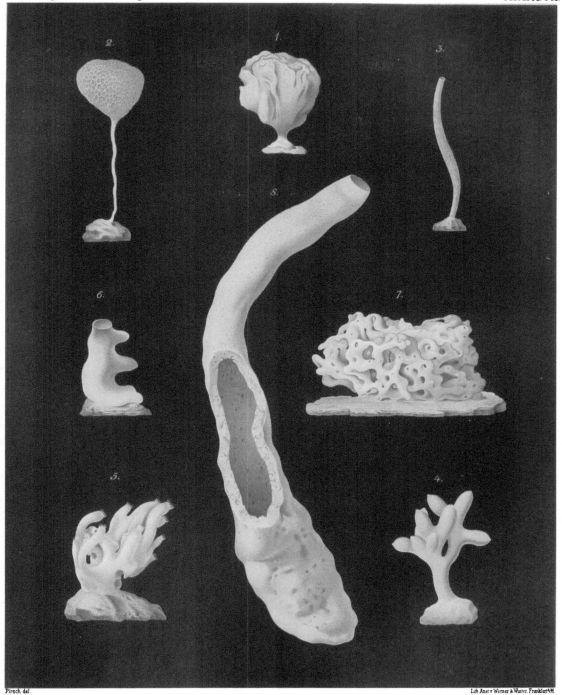

1. Leucosolenia Challengeri.
2. Leucosolenia blanca.
3. Ute argentea.
4. Sycon arboreum.
5. Amphoriscus flamma.
6. Grantia tuberosa.
7. Heteropegma nodus Gordii.
8. Leuconia multiformis var. Goliath.

The Voyage of H.M.S "Challenger." Calcarea Pl.I

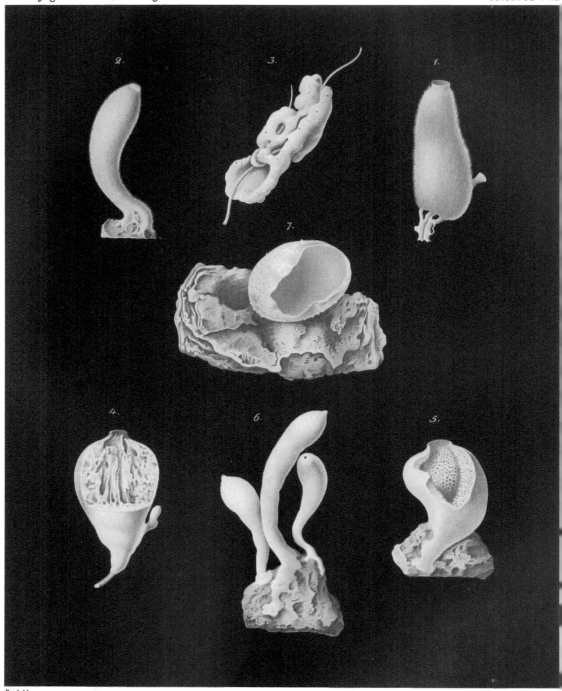

1. *Leuconia multiformis* var. *capillata*.
2. *Leuconia loricata*.
3. *Leuconia dura*.
4. *Leuconia fruticosa*.
5. *Pericharax Carteri*.
6. *Leucetta Haeckeliana*.
7. *Eilhardia Schulzei*.

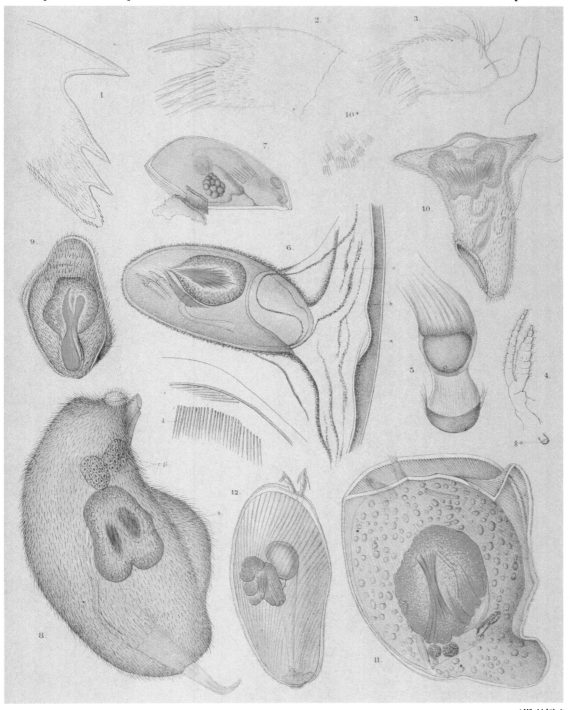

1-5 SCALPELLUM PARALLELOGRAMMA, Hoek. 6 SC. NYMPHOCOLA, Hoek. 7-9 SC. VELUTINUM, Hoek. 10 SC. EXIMIUM, Hoek. 11 SC. GIGAS, Hoek. 12 SC. REGIUM, (W. Th.) Hoek.

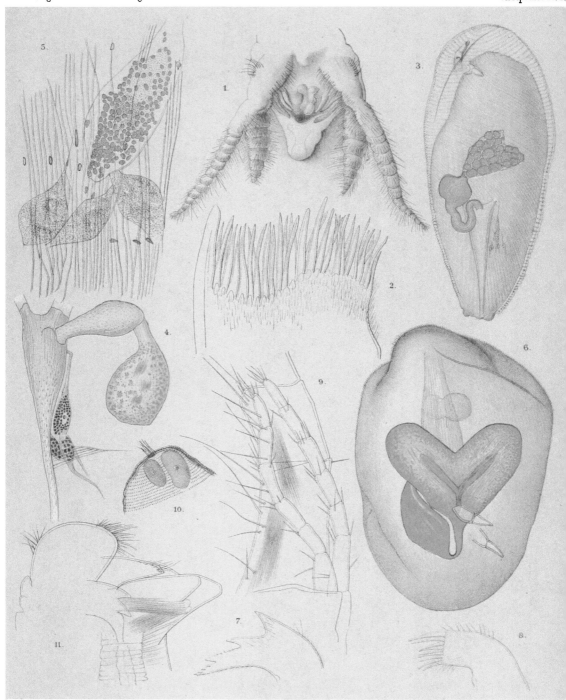

1-2 SCALPELLUM REGIUM, (Wyv. Thoms.) Hoek. 3-5 SC. DARWINII, Hoek. 6 SC. TENUE. Hoek.
7-9 SC. FLAVUM, Hoek. 10 SC. TRITONIS, Hoek. 11 SC. BALANOIDES, Hoek.

第九卷动物志九(图版)彩图

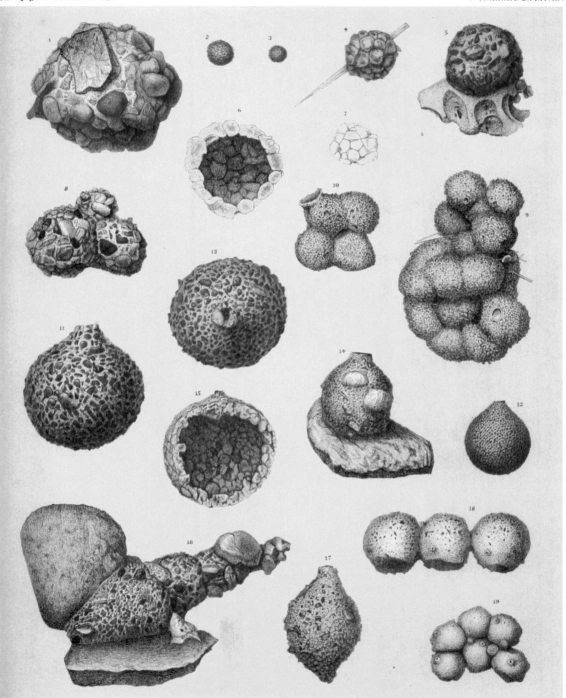

PSAMMOSPHÆRA—SOROSPHÆRA—SACCAMMINA.

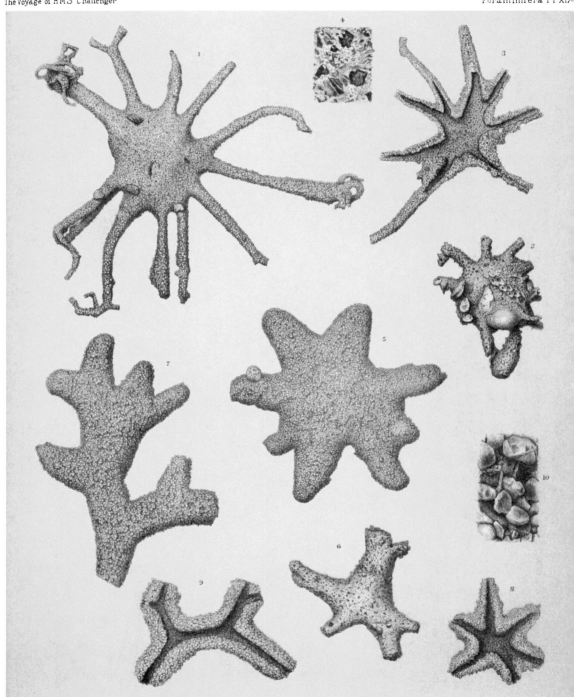

ASTRORHIZA.

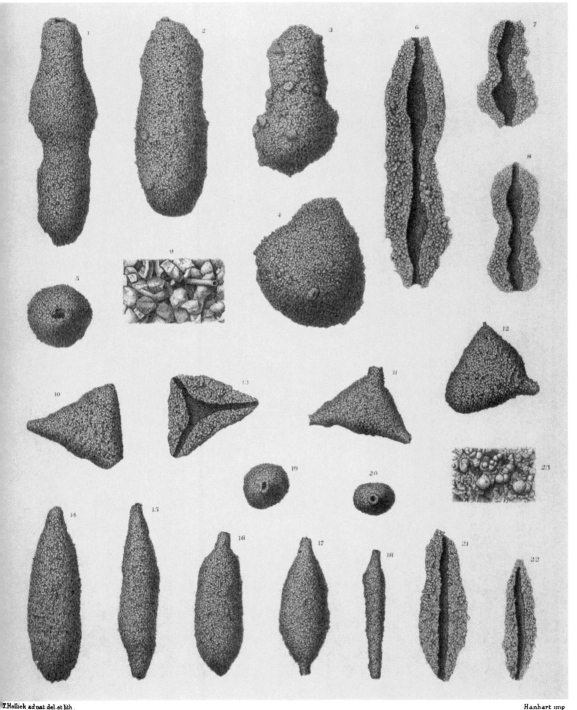

ASTRORHIZA.

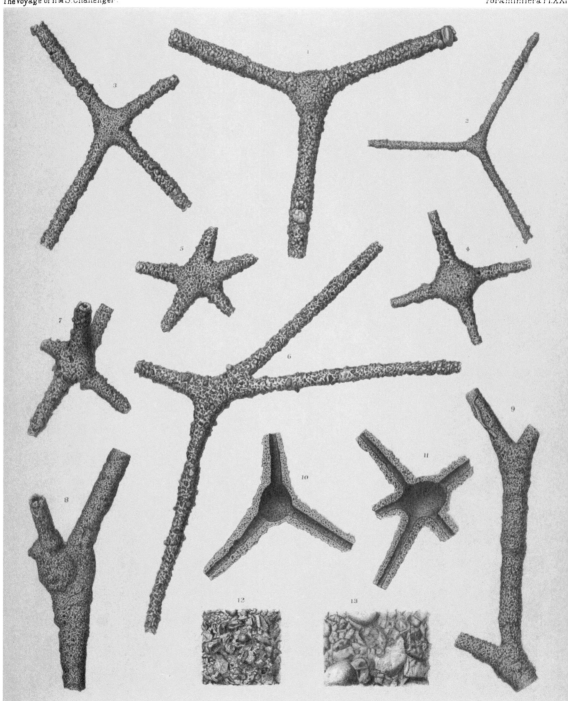

RHABDAMMINA.

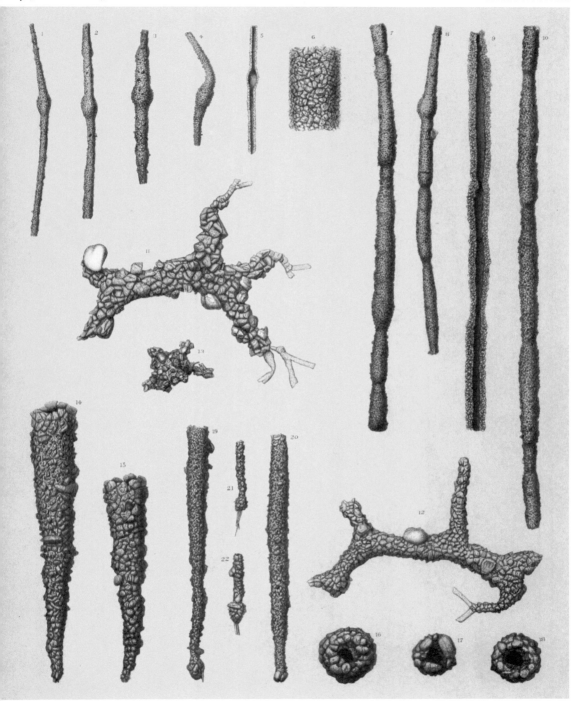

RHABDAMMINA_JACULELLA.

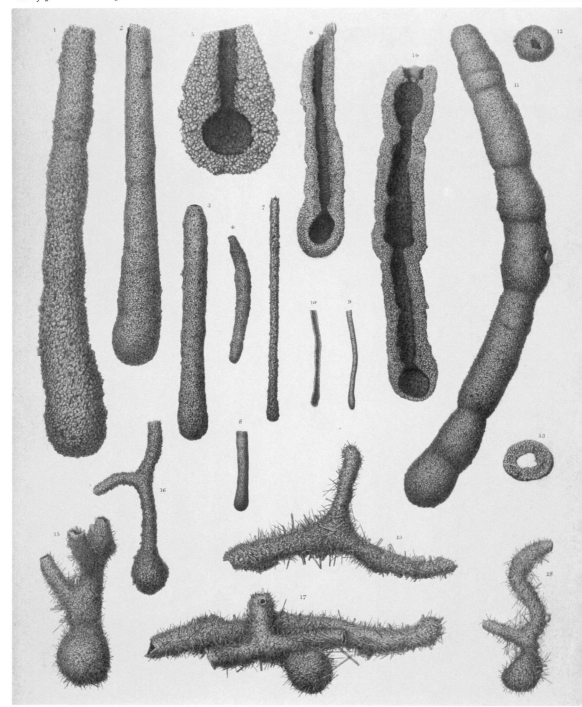

HYPERAMMINA.

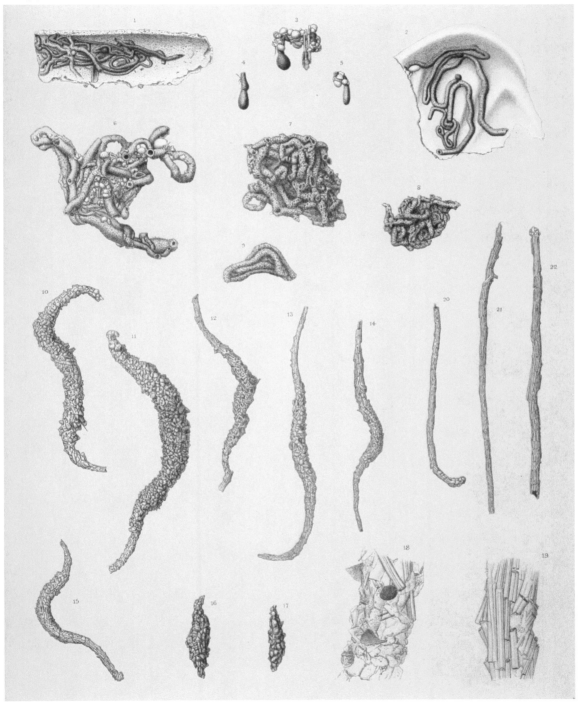

HYPERAMMINA — MARSIPELLA

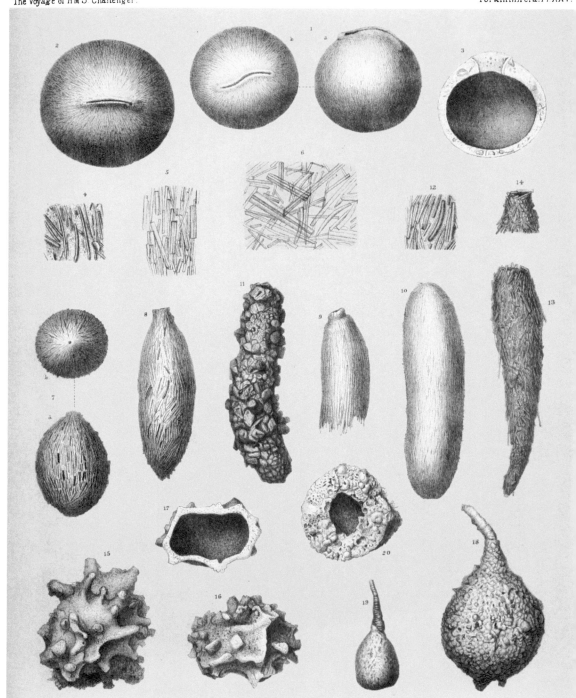

PILULINA—TECHNITELLA—STORTHOSPHÆRA—PELOSINA.

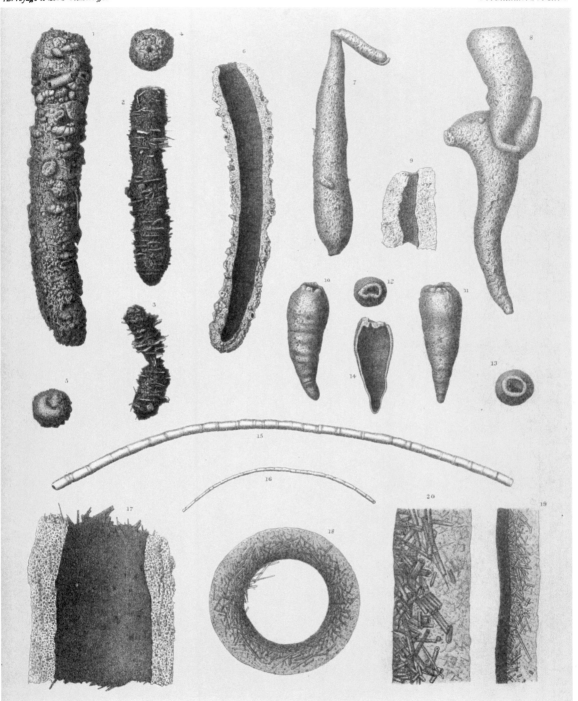

PELOSINA – HIPPOCREPINA – BATHYSIPHON.

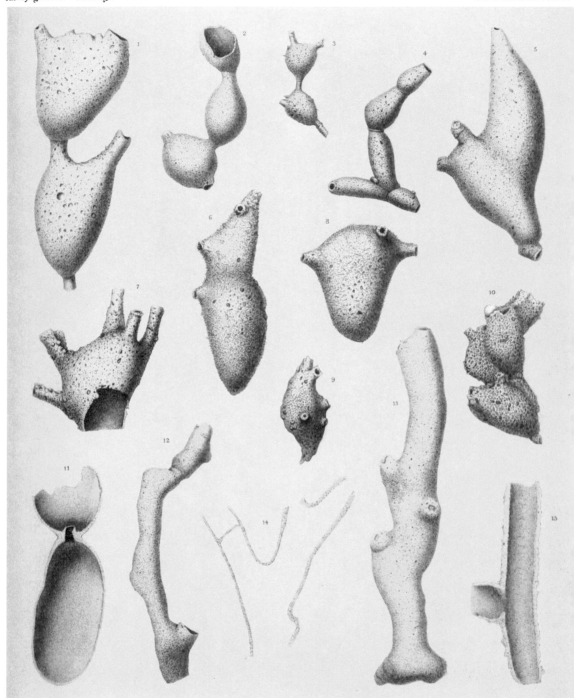

ASCHEMONELLA.

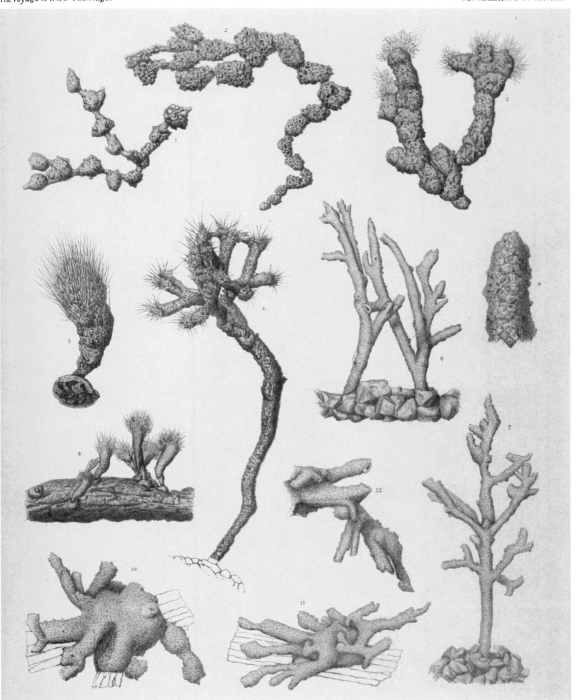

ASCHEMONELLA _ HALIPHYSEMA _ DENDROPHRYA.

The Voyage of H.M.S. "Challenger" — Foraminifera Pl. XXVI

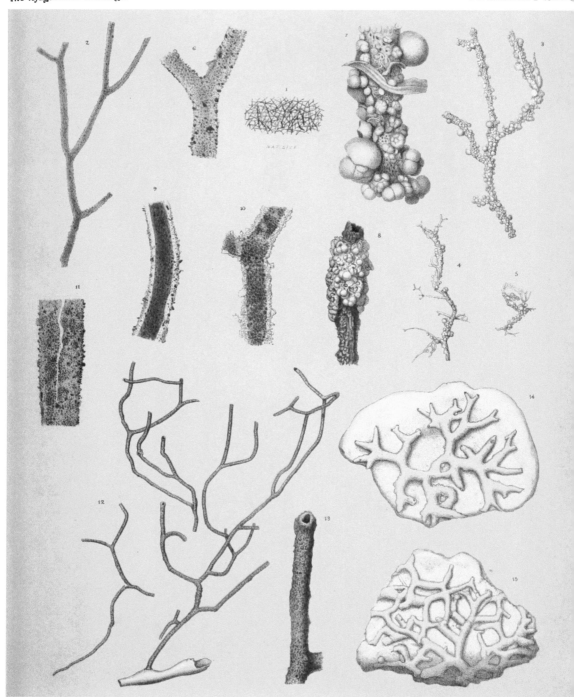

A.T. Hollick ad nat. del. et lith. Hanhart imp.

RHIZAMMINA_HYPERAMMINA_SAGENELLA.

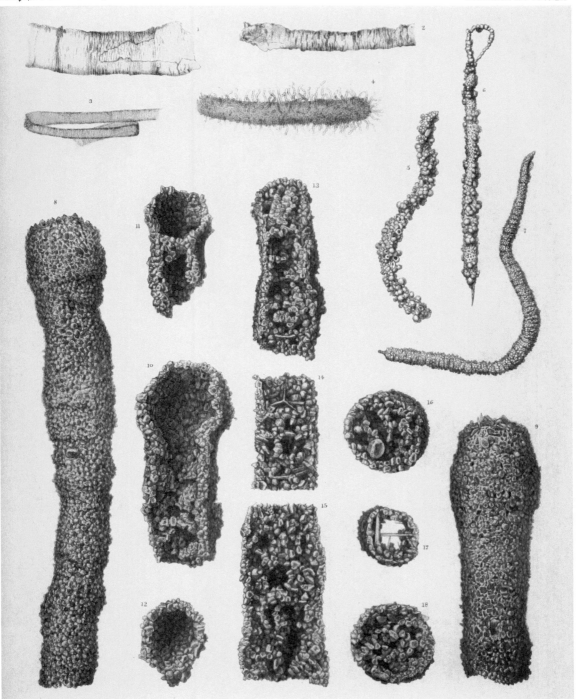

RHIZAMMINA_BOTELLINA.

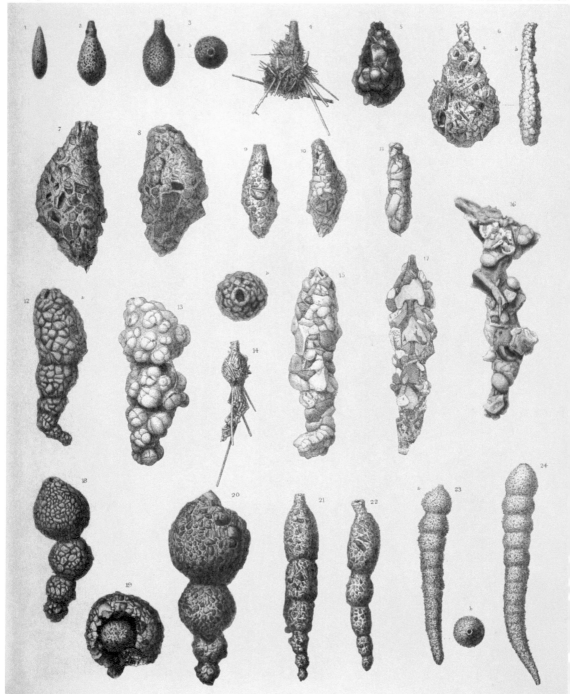

LITUOLA (Reophax)

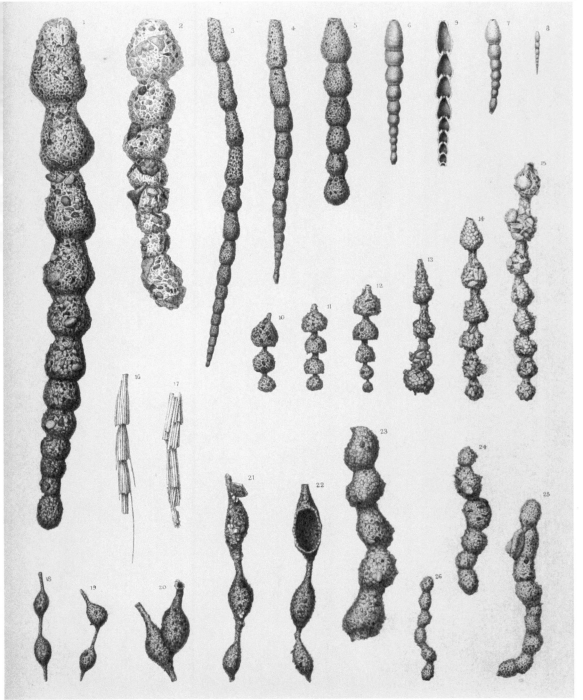

LITUOLA (Reophax)

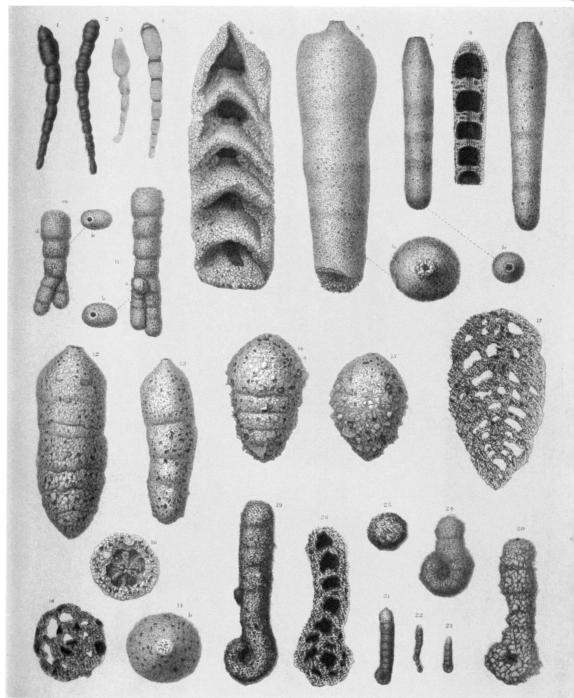

LITUOLA. (Reophax. Haplostiche. Haplophragmium.)

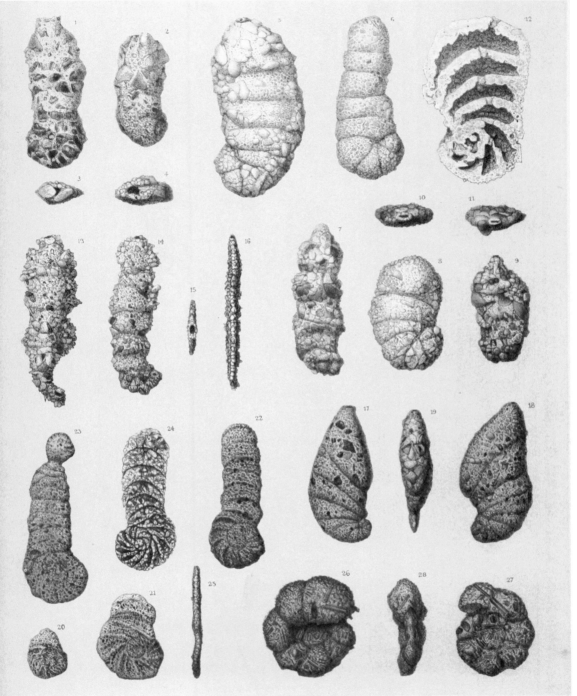

LITUOLA (Haplophragmium)

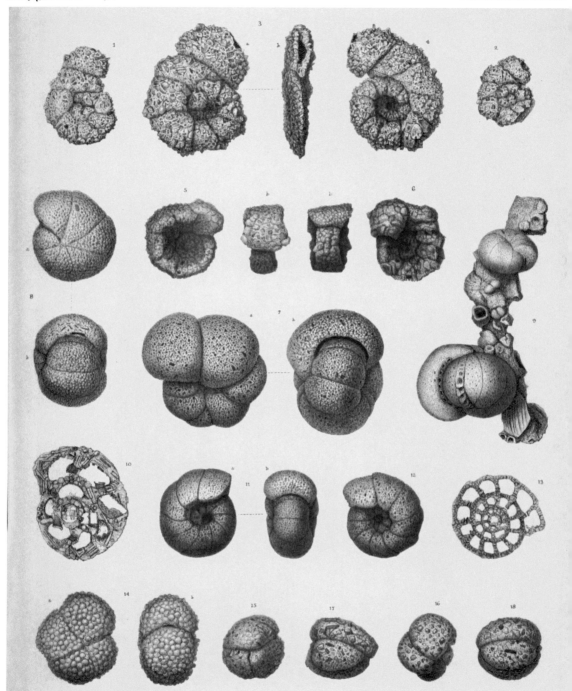

LITUOLA, (Haplophragmium.)

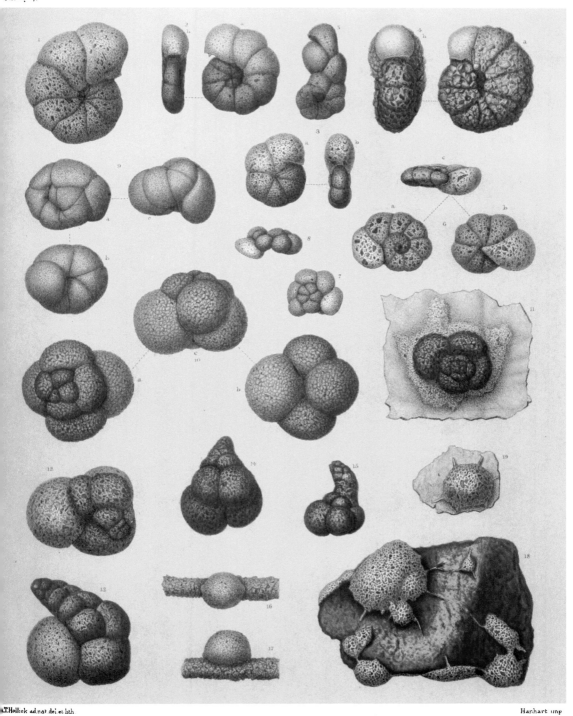

LITUOLA. (Haplophragmium_Placopsilina.)

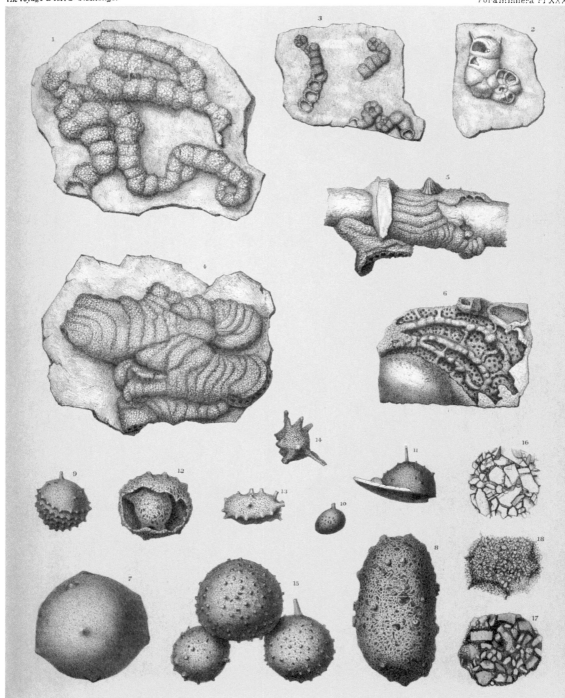

LITUOLA (Placopsilina-Bdelloidina)-THURAMMINA.

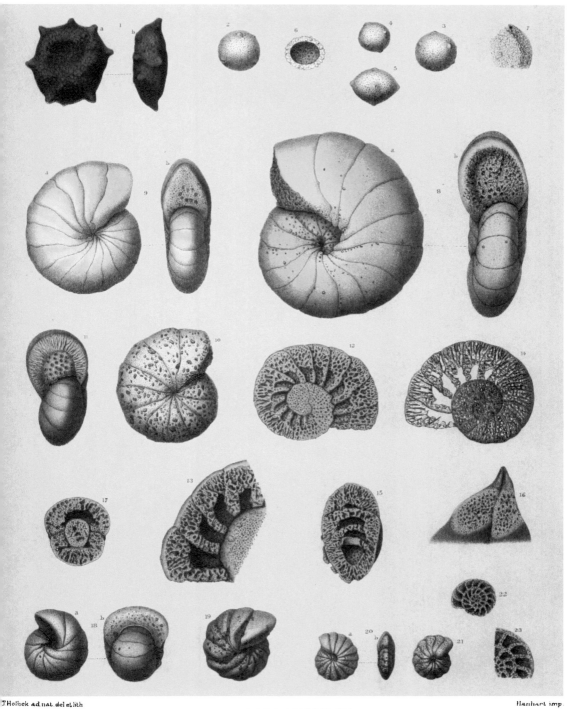

THURAMMINA_CYCLAMMINA.

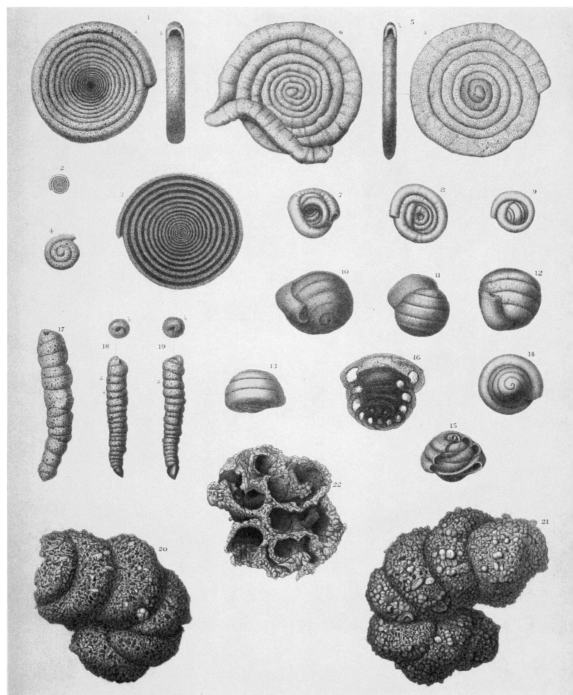

TROCHAMMINA.(Ammodiscus)

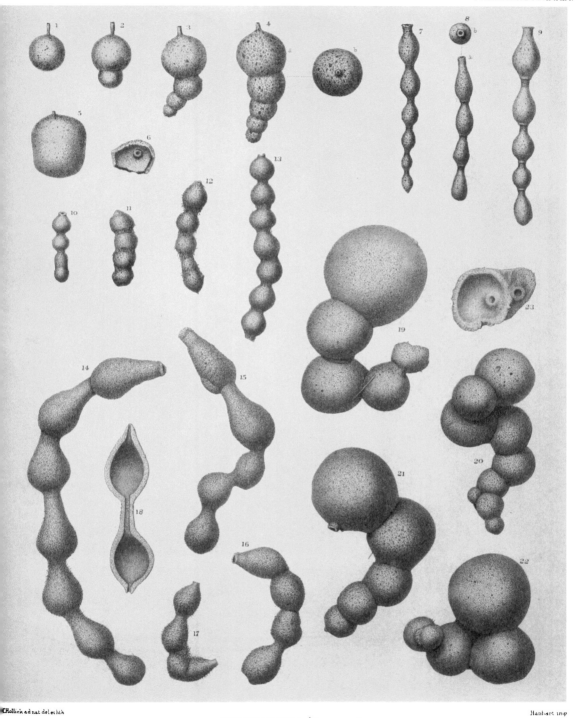

TROCHAMMINA. (Hormosina)

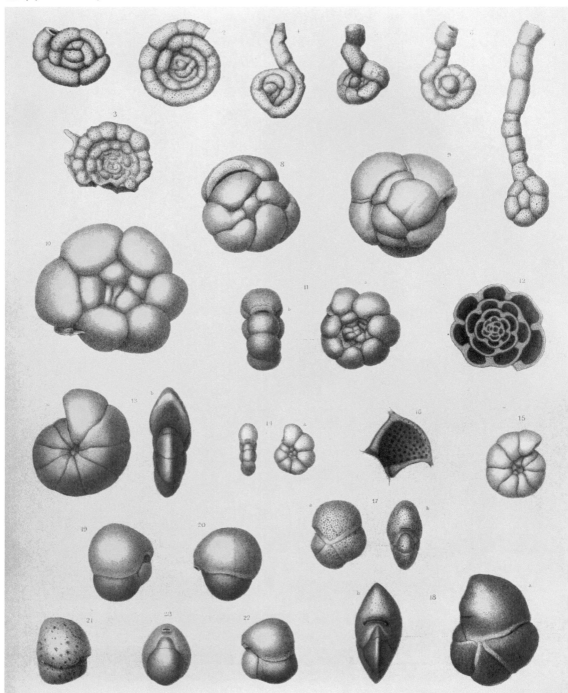

TROCHAMMINA.

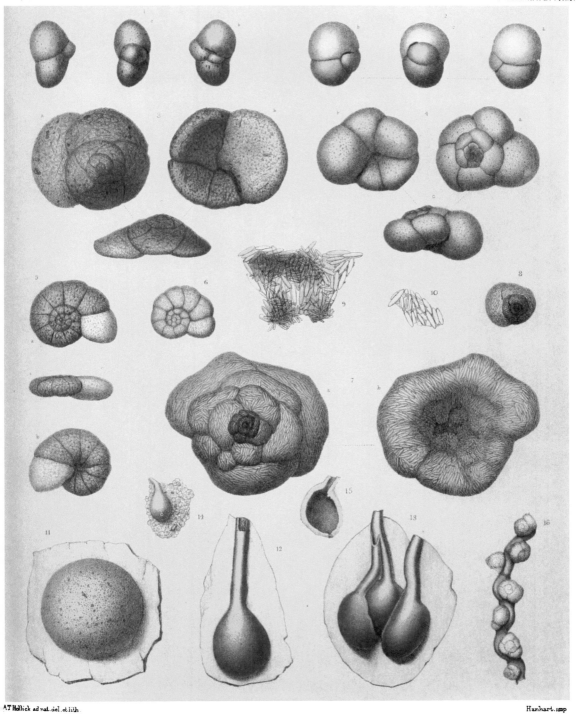

TROCHAMMINA. (Trochammina_Carterina_Webbina.)

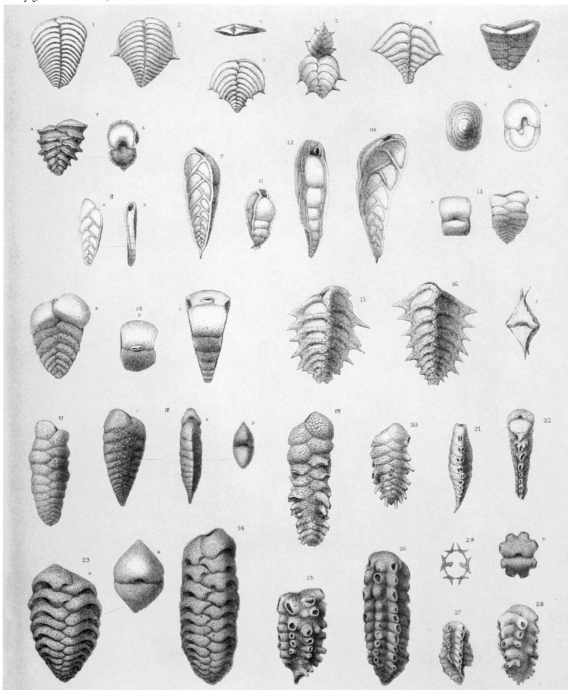

TEXTULARIA.

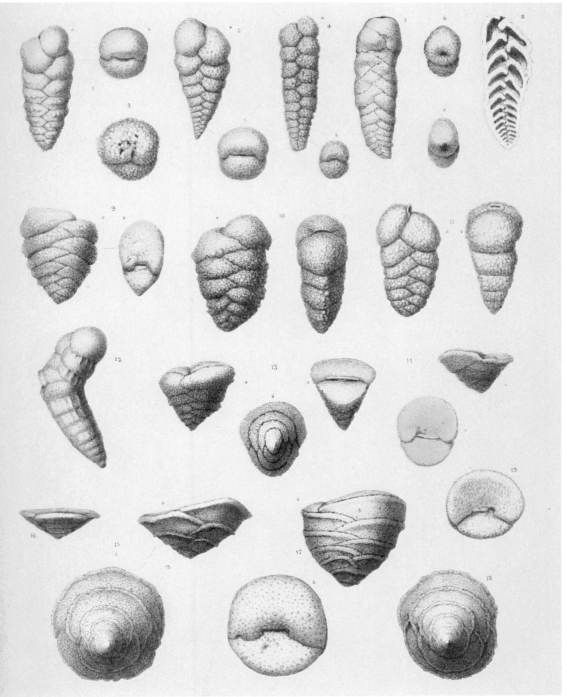

TEXTULARIA

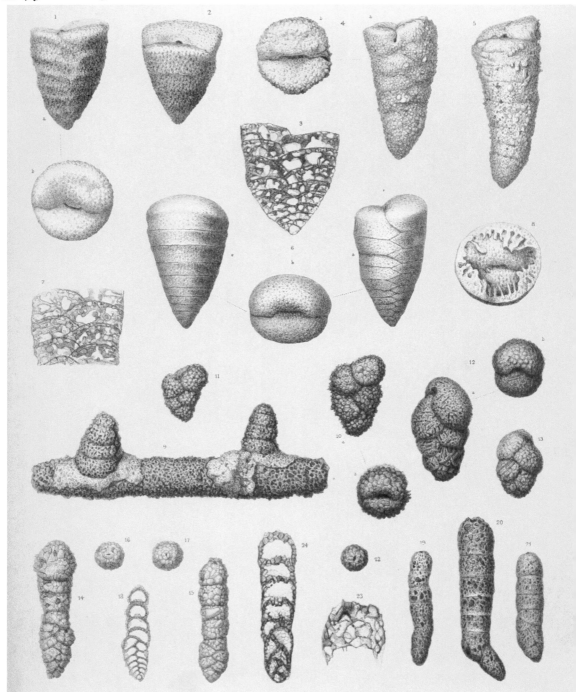

TEXTULARIA (Textularia_Bigenerina.)

TEXTULARIA (Bigenerina_Pavonina_Spiroplecta)

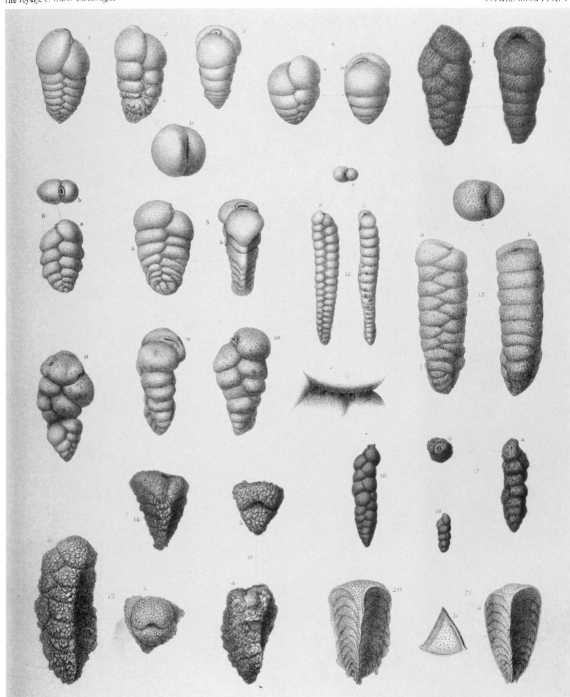

TEXTULARIA, (Gaudryina _ Chrysalidina.)

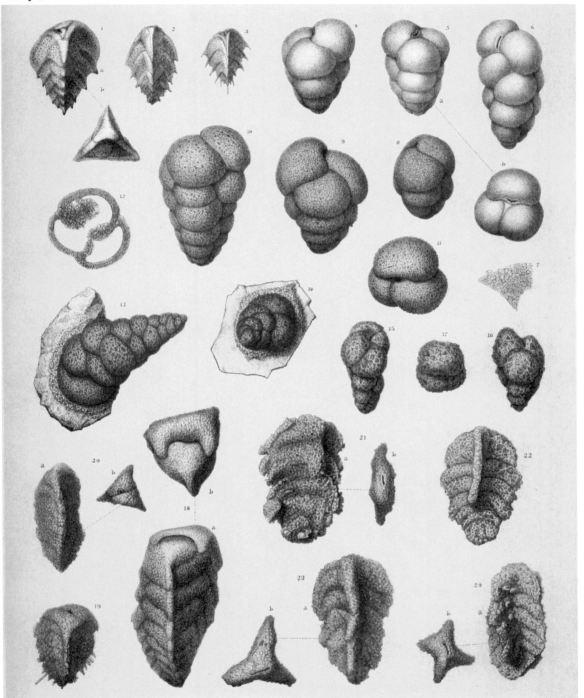

TEXTULARIA (Verneuilina.)

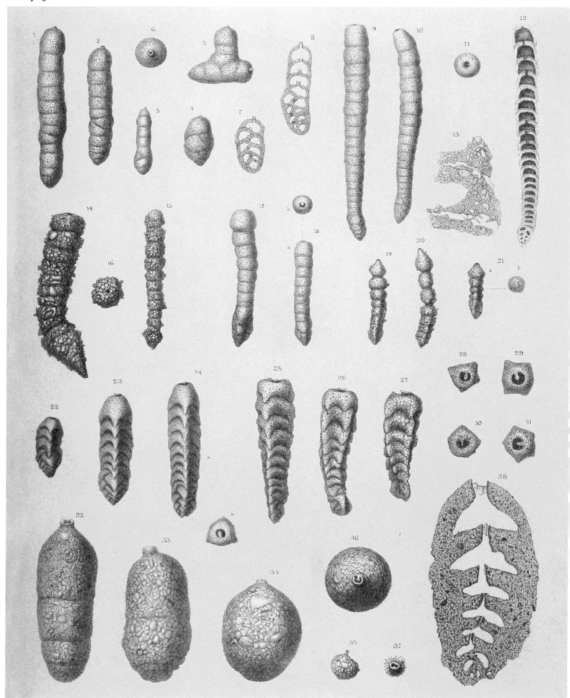

VALVULINA (Clavulina)

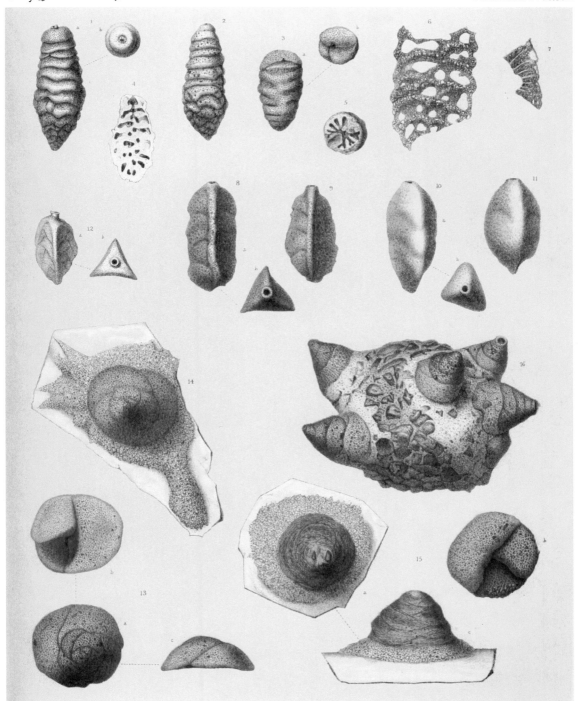

TEXTULARIA (Tritaxia). VALVULINA.

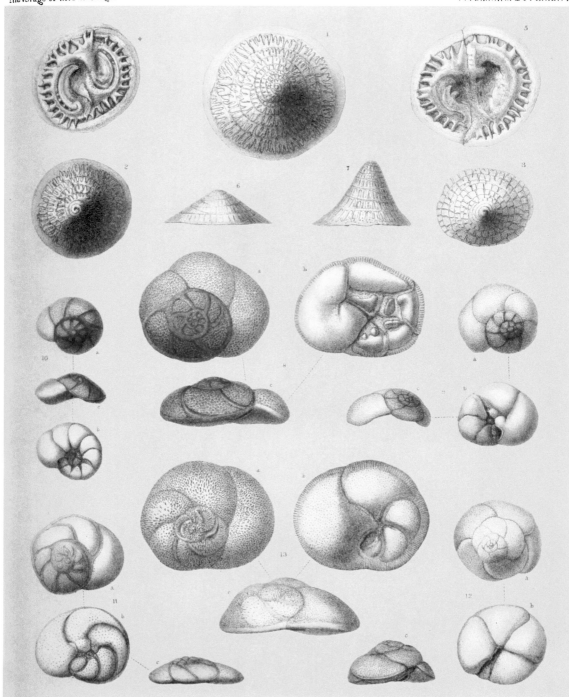

PATELLINA — DISCORBINA.

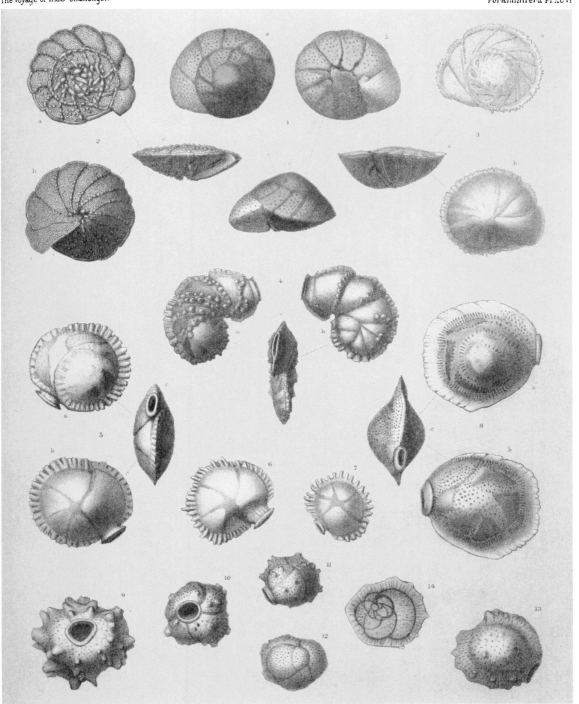

PLANORBULINA (Truncatulina).

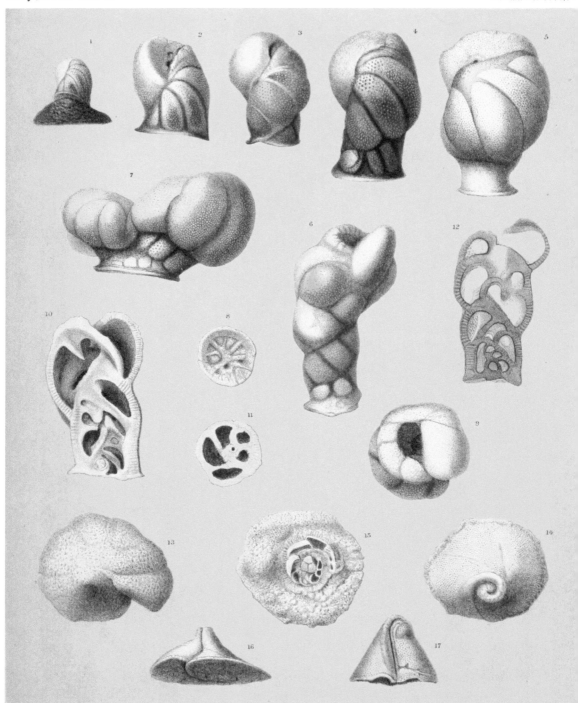

RUPERTIA — CARPENTERIA.

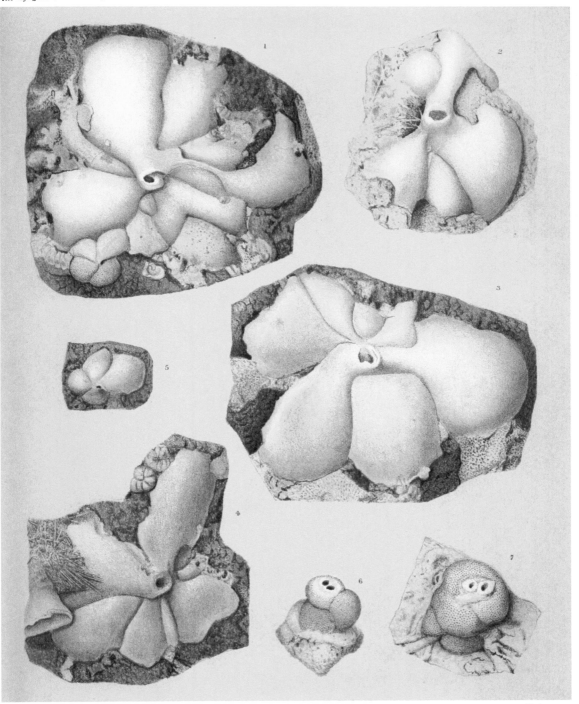

CARPENTERIA.

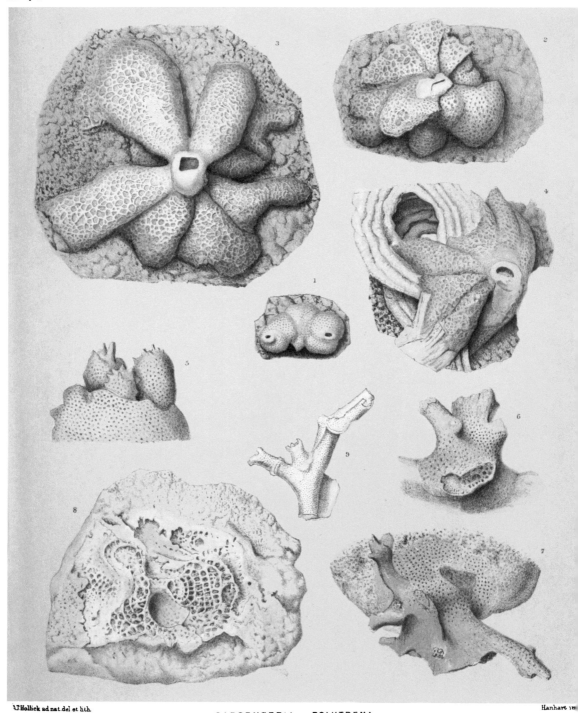

CARPENTERIA — POLYTREMA.

第十一卷动物志十彩图

The Voyage of H.M.S. "Challenger." Myzostomida Pl.I

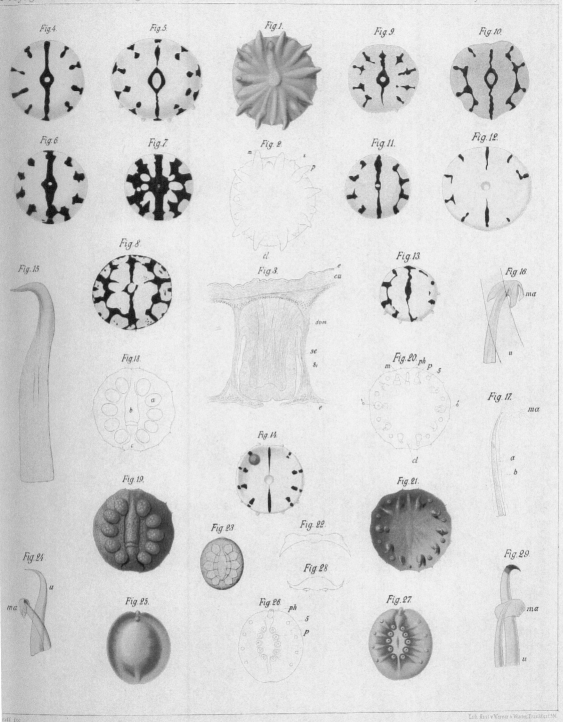

1-17. Myzostoma horologium. 18-24. M. testudo. 25-29. M. alatum.

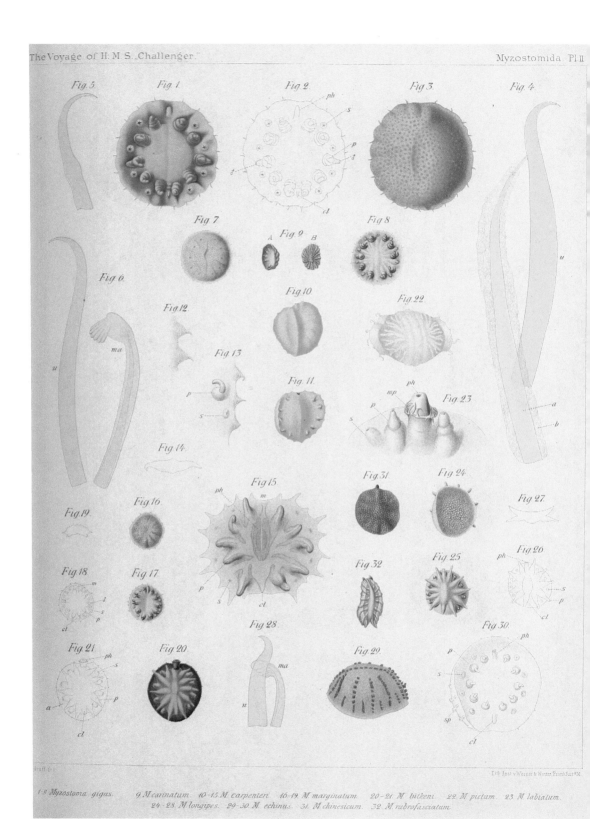

1-8. *Myzostoma gigas.* 9. *M. carinatum.* 10-15. *M. carpenteri.* 16-19. *M. marginatum.* 20-21. *M. lütkeni.* 22. *M. pictum.* 23. *M. labiatum.* 24-28. *M. longipes.* 29-30. *M. echinus.* 31. *M. chinesicum.* 32. *M. rubrofasciatum.*

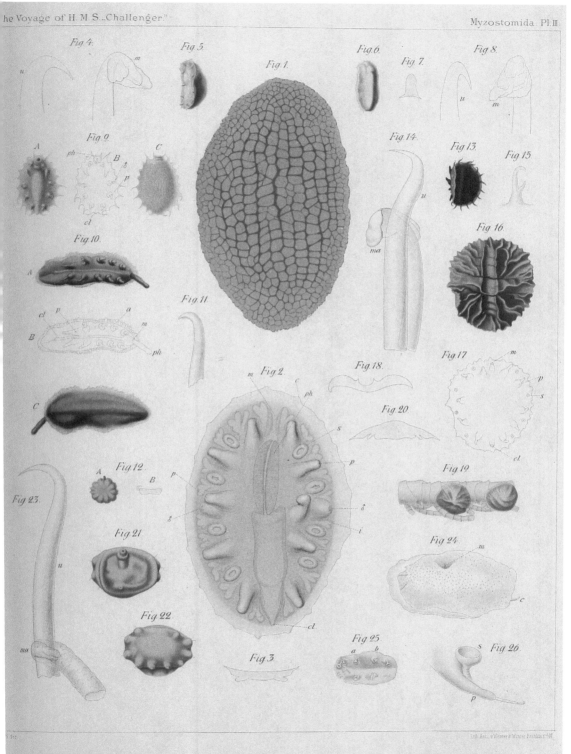

1-3. Myzostoma areolatum. 4-8. M. compressum. 9. M. coronatum. 10-11. M. folium. 12. M. radiatum. 13-15. M. nigrescens. 16-18. M. plicatum. 19-20. M. brevipes. 21-23. M. pulvinar. 24-26. M. calycocotyle.

1. *Myzostoma fissum.* 2. *M. intermedium.* 3-6. *M. quadrifilum.*

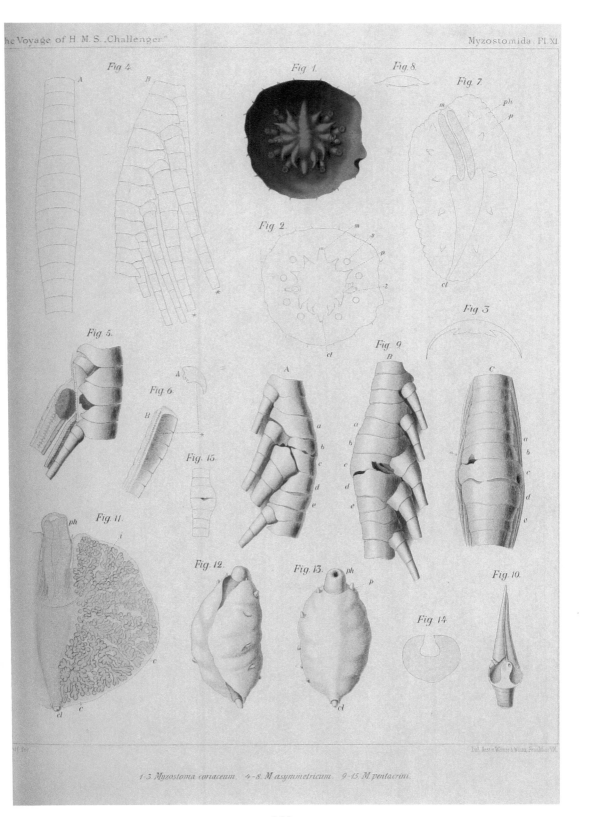

1-3. *Myzostoma coriaceum.* 4-8. *M. asymmetricum.* 9-15. *M. pentacrini.*

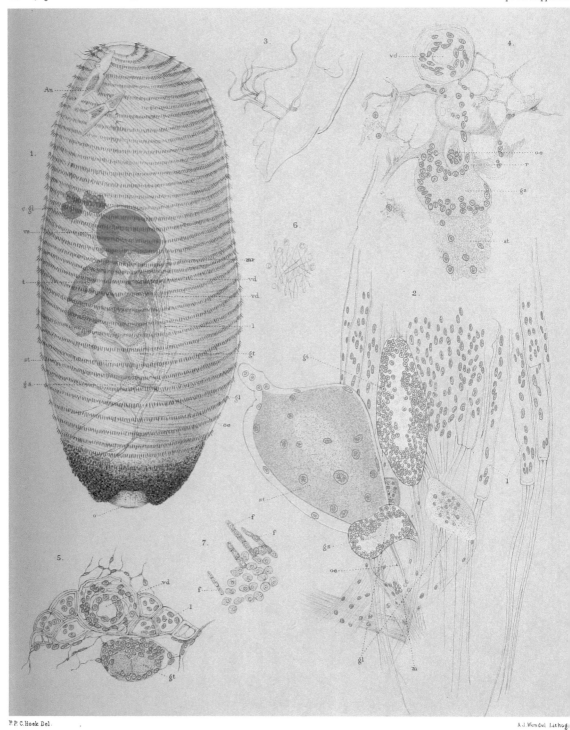

MALE OF SCALPELLUM REGIUM.

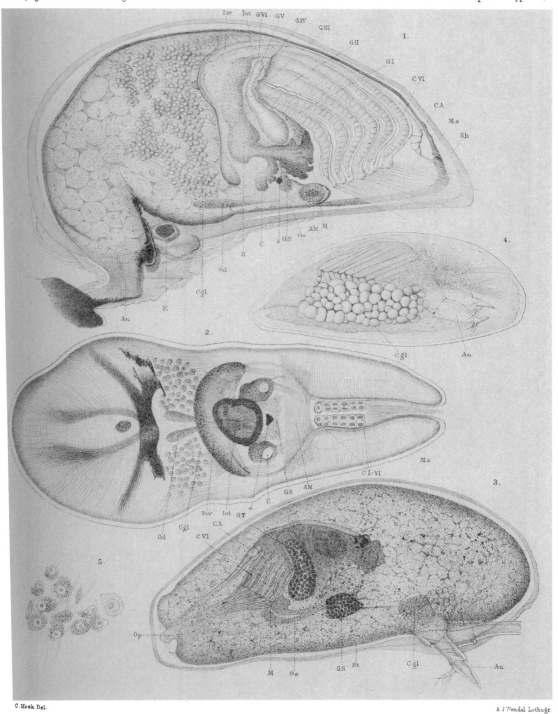

CYPRIS-LARVAE OF CIRRIPEDIA.

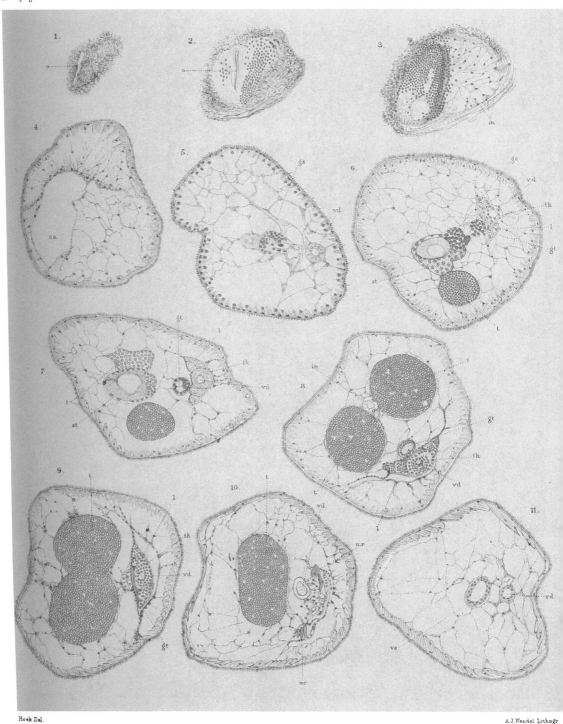

MALE OF SCALPELLUM REGIUM.

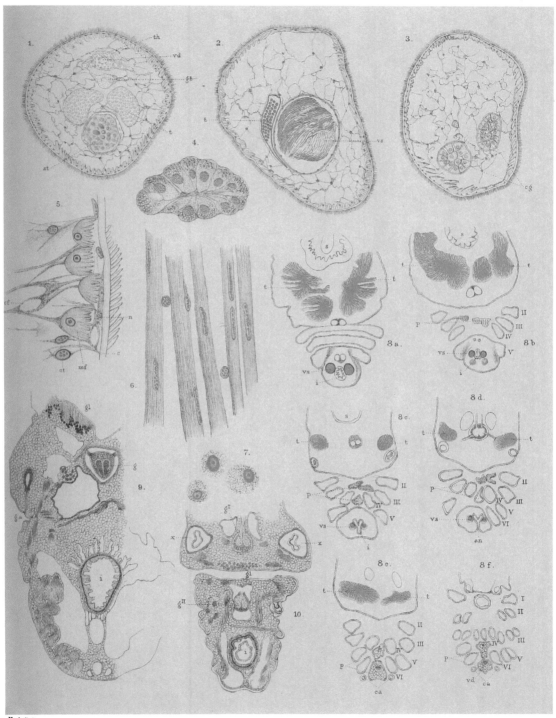

1-7 MALE OF SC. REGIUM. 8 SC. BALANOIDES. 9 SC. PARALLELOGRAMMA. 10 SC. NYMPHOCOLA.

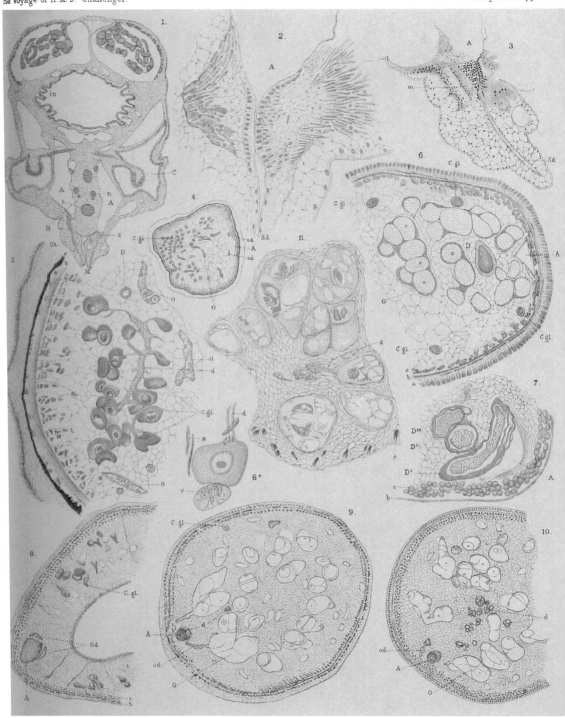

1-3 SCALPELLUM VULGARE. 4-5 LEPAS ANATIFERA. 6-7 SC. VULGARE. 8-11 SC. REGIUM.

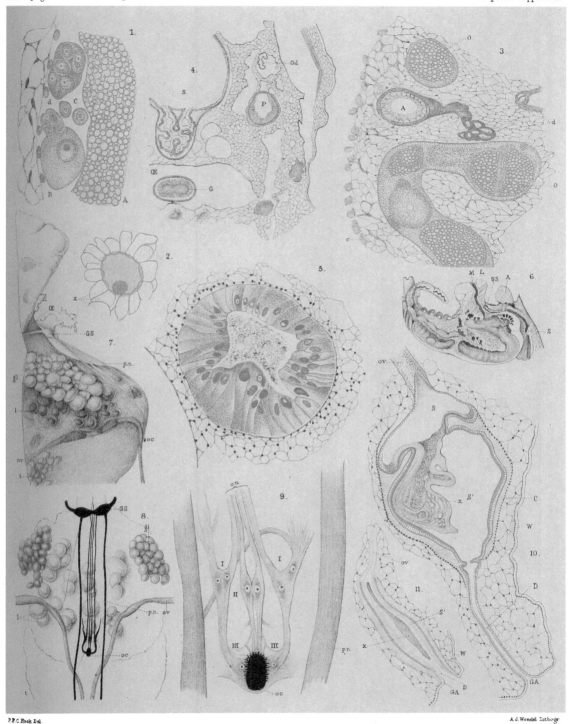

1-2 SCALPELLUM VULGARE. 3 SC. REGIUM. 4-5 SC. PARALLELOGRAMMA.
6-9 LEPAS ANATIFERA. 10 SC. VULGARE. 11 LEP. HILLII.

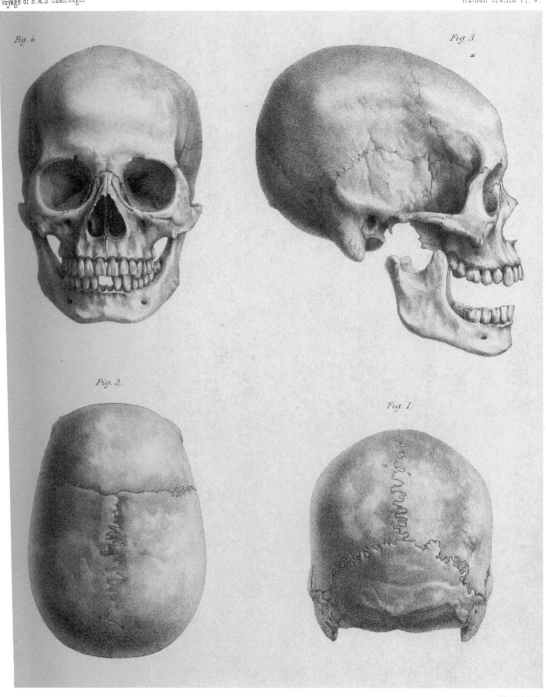

FIGS. 1 & 2, ADMIRALTY ISLANDER. FIGS. 3 & 4, SANDWICH ISLANDER, (Waimea)

第十二卷动物志十一彩图

PENTACRINUS WYVILLE—THOMSONI, Jeffreys.

PENTACRINUS WYVILLE-THOMSONI, Jeffreys.

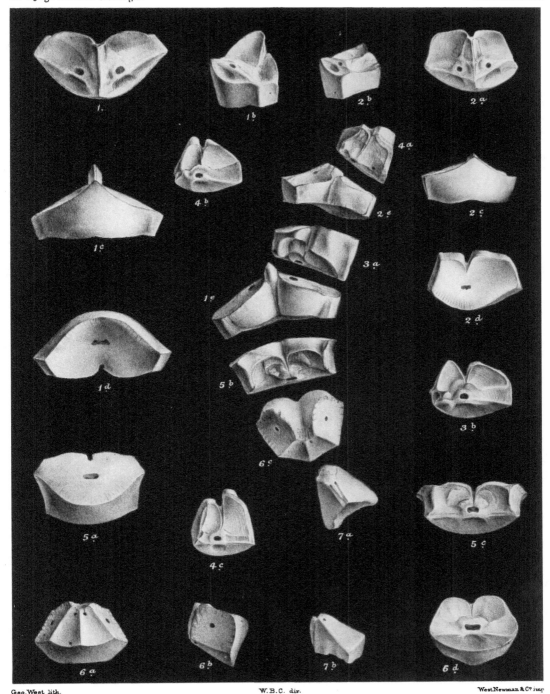

PENTACRINUS WYVILLE-THOMSONI, Jeffreys.

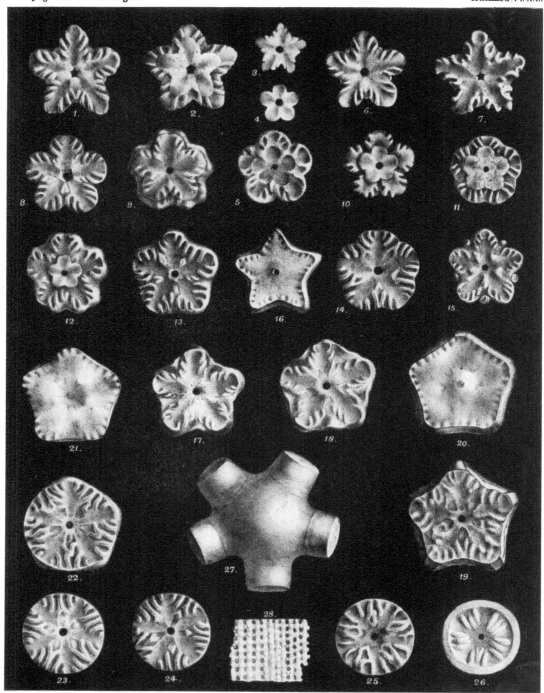

PENTACRINUS, WYVILLE—THOMSONI, Jeffreys

PENTACRINUS WYVILLE-THOMSONI, Jeffreys.

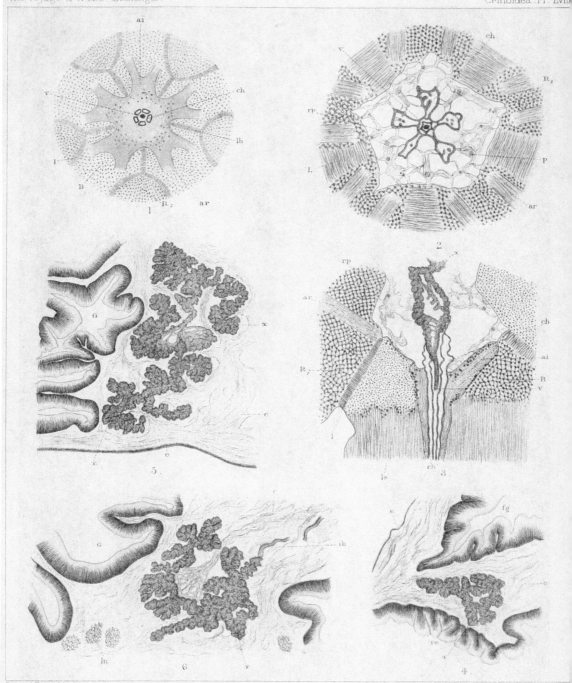

1–3. PENTACRINUS DECORUS, Wyv Th. 4. PENTACRINUS NARESIANUS, Sp.n.
5, 6. PROMACHOCRINUS KERGUELENSIS, Sp.n.

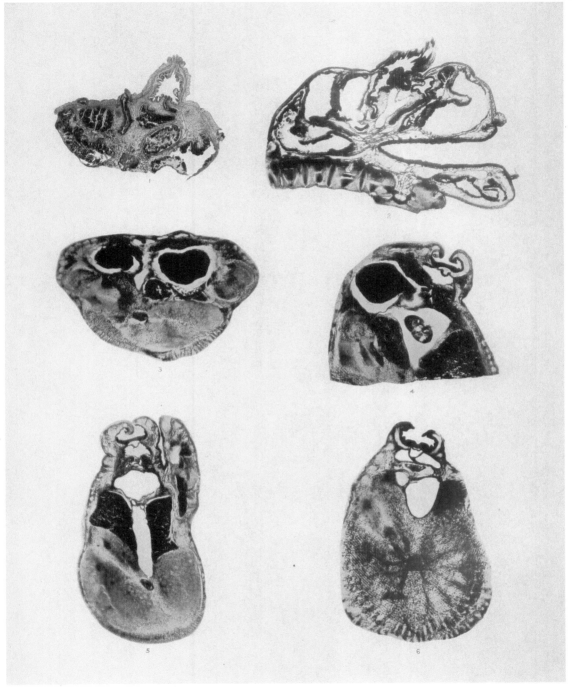

1. ACTINOMETRA PULCHELLA, Pourt., Sp.   2-5. ACTINOMETRA PARVICIRRA, Mull., Sp
6. ACTINOMETRA NIGRA, Sp. n.

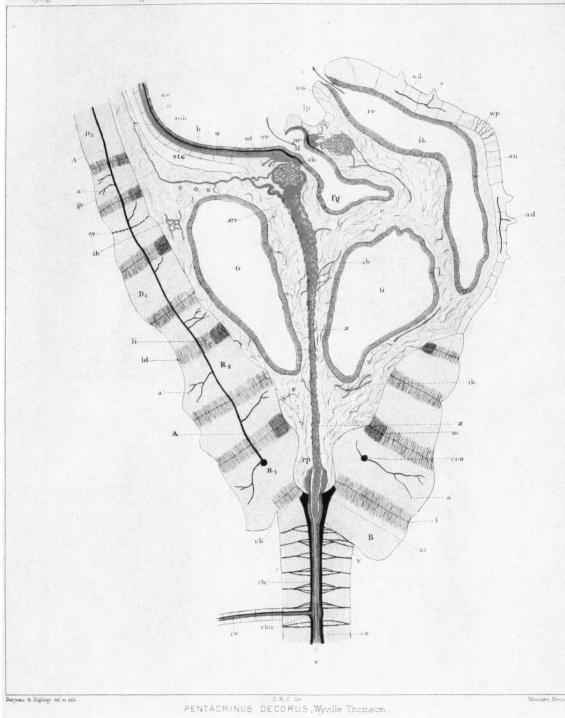

PENTACRINUS DECORUS, Wyville Thomson.

第十三卷动物志十二彩图

AMPHINOMIDÆ & POLYNOIDÆ.

ALCIOPIDÆ.

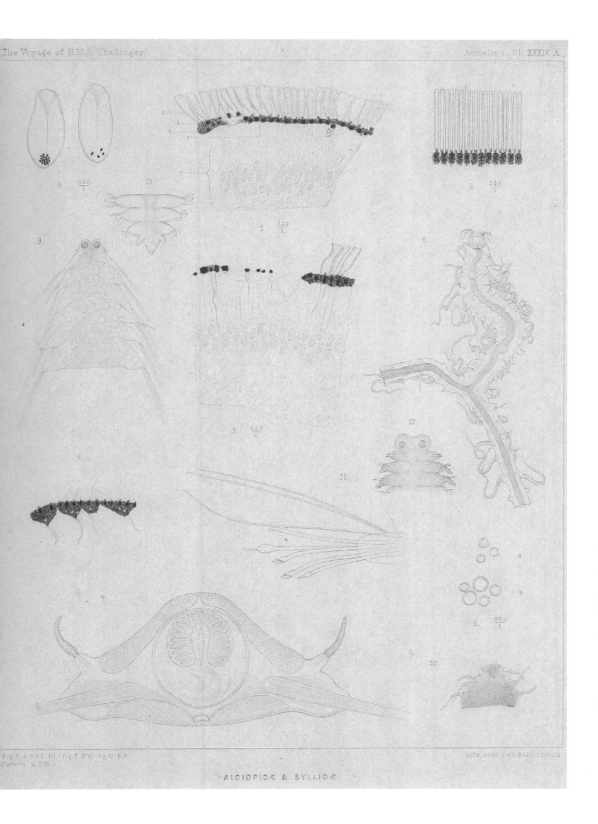

ALCIOPIDÆ & SYLLIDÆ.

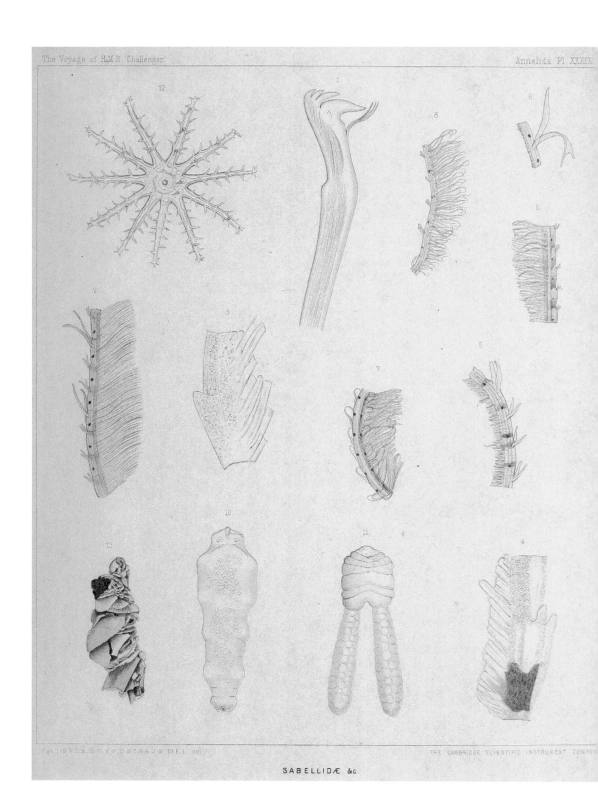

SABELLIDÆ &c.

第十五卷动物志十四彩图

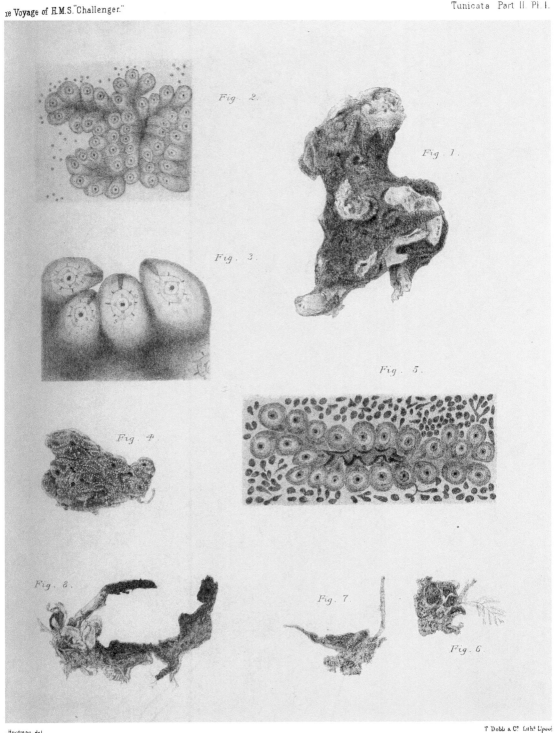

FIGS 1-3 BOTRYLLOIDES PURPUREUM, n. sp.
FIGS 4-5 BOTRYLLOIDES PERSPICUUM, n. sp.
FIGS 6-7 BOTRYLLOIDES PERSPICUUM, var. RUBICUNDUM n.
FIG 8   BOTRYLLOIDES NIGRUM, n. sp.

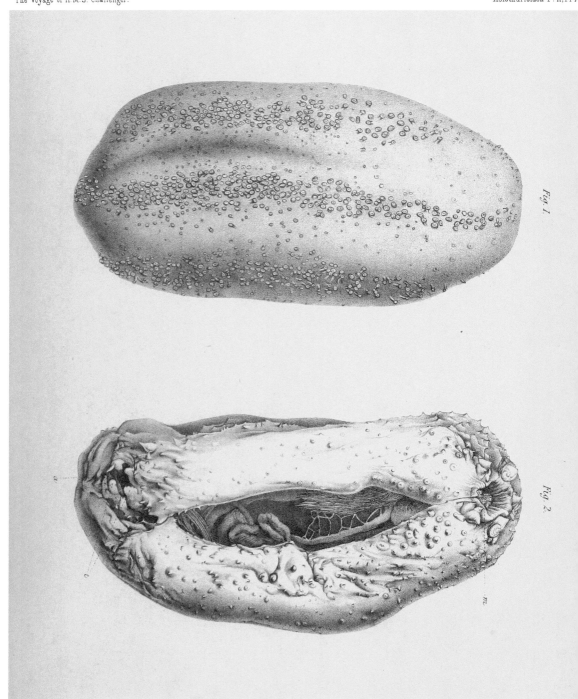

COLOCHIRUS VIOLACEUS, n. sp.

第十六卷动物志十五彩图

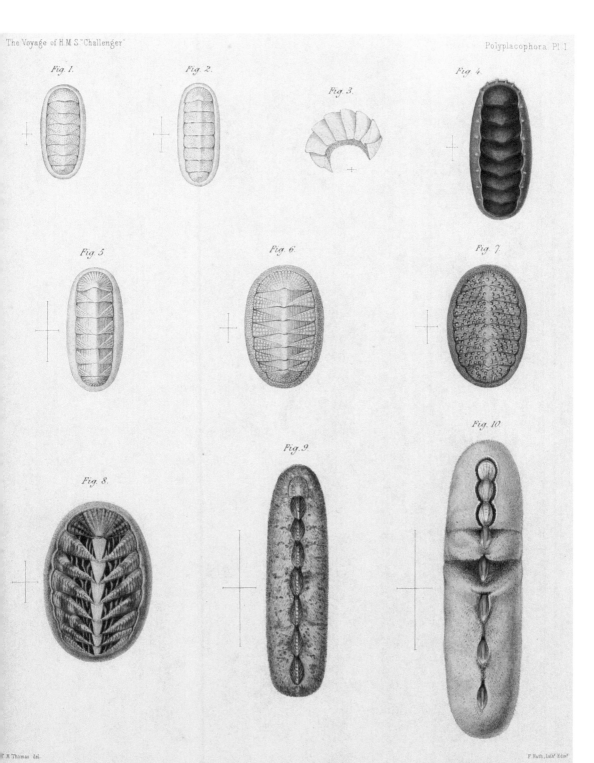

1, LEPTOCHITON BENTHUS, n.sp.   2, LEPTOCHITON BELKNAPI, Dall.   3, LEPTOCHITON KERGUELENSIS, n.sp.
4, HEMIARTHRUM SETULOSUM, Carpenter.   5, LEPIDOPLEURUS DORSUOSUS, n.sp.   6, LEPIDOPLEURUS DALLII, n.sp.
7, CHITON MURRAYI, n.sp.   8, PLAXIPHORA CARPENTERI, n.sp.   9, CRYPTOPLAX STRIATUS, Lamarck.
10, CRYPTOPLAX OCULATUS, Quoy & Gaimard.

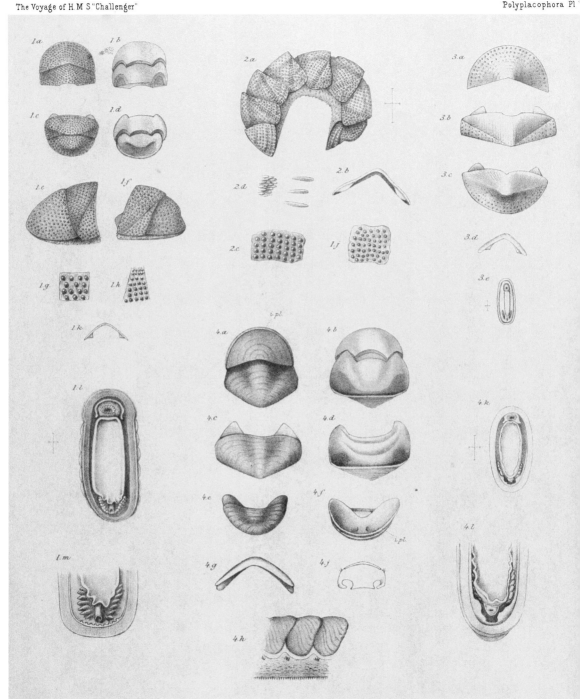

1a–1m, LEPTOCHITON BENTHUS, n.sp. 2a–2d, LEPTOCHITON BELKNAPI, Dall.
3a–3c, LEPTOCHITON KERGUELENSIS, n.sp. 4a–4l, HEMIARTHRUM SETULOSUM, Carpenter.

第十七卷动物志十六彩图

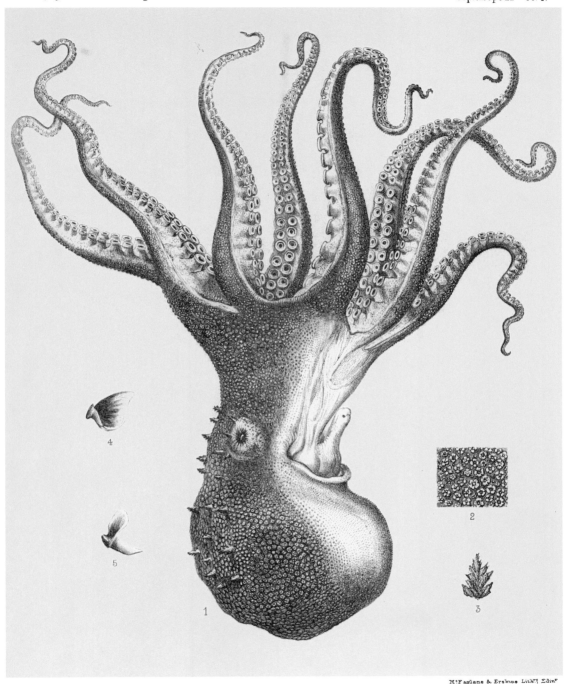

OCTOPUS BOSCII var. PALLIDA, nov.

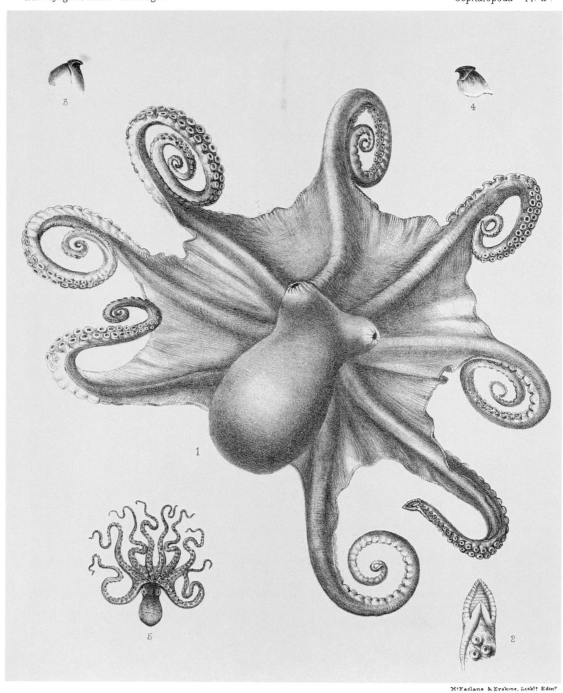

1-4, OCTOPUS LEVIS, n sp.   5, OCTOPUS BERMUDENSIS, n. sp.

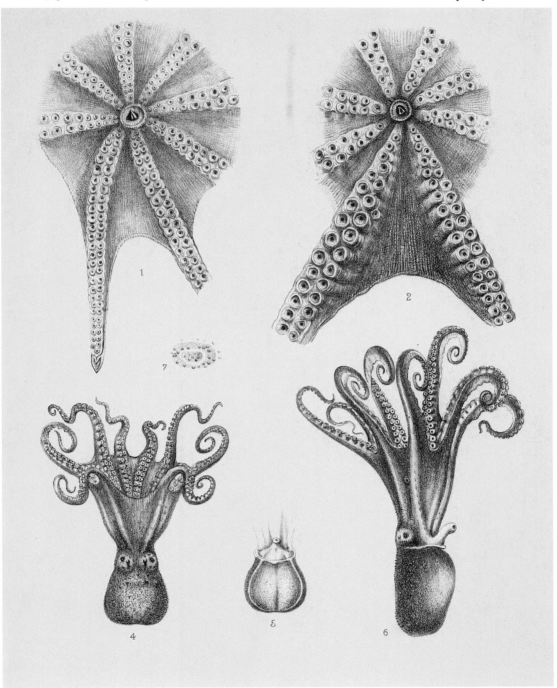

1, OCTOPUS LEVIS, n. sp. 2, OCTOPUS BOSCII, var. PALLIDA nov.
4, 5, OCTOPUS AUSTRALIS, n. sp. 6, 7, OCTOPUS AREOLATUS, de Haan.

OCTOPUS VERRUCOSUS, n.sp.

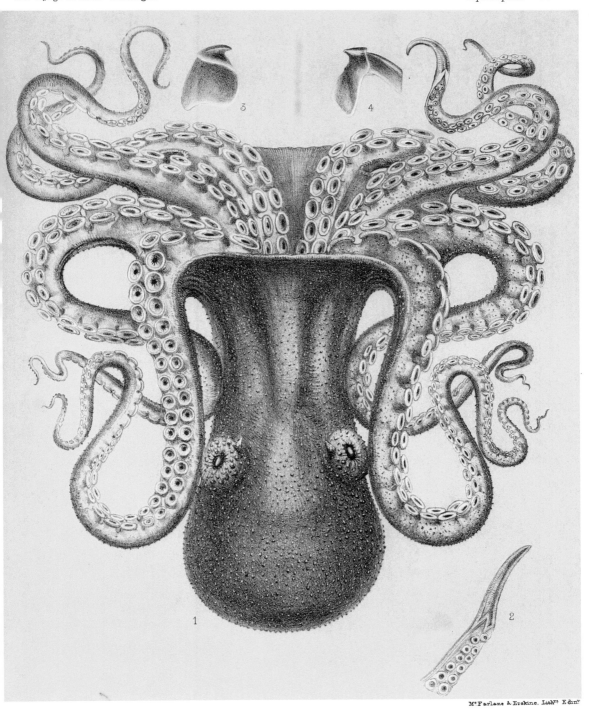

OCTOPUS HONGKONGENSIS, Stp.

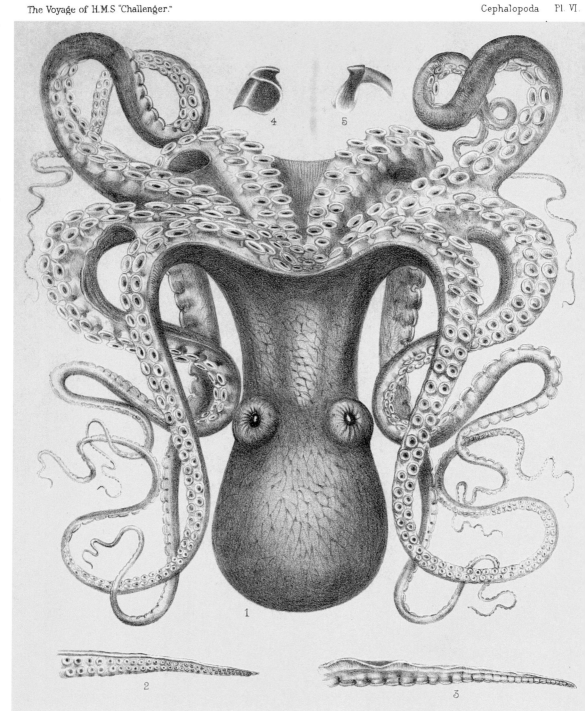

OCTOPUS MARMORATUS, n. sp.

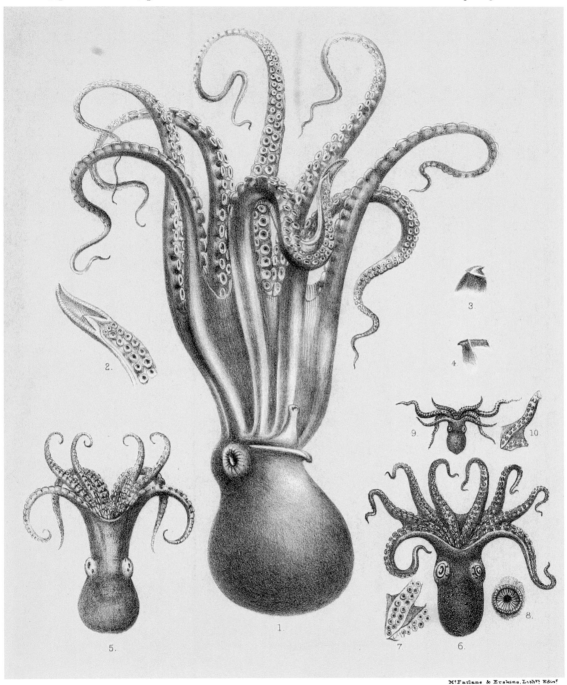

1-4, OCTOPUS JANUARII, Stp.   5, OCTOPUS DUPLEX, n. sp.
6-8, OCTOPUS VITIENSIS, n. sp. 9,10, OCTOPUS BANDENSIS, n. sp.

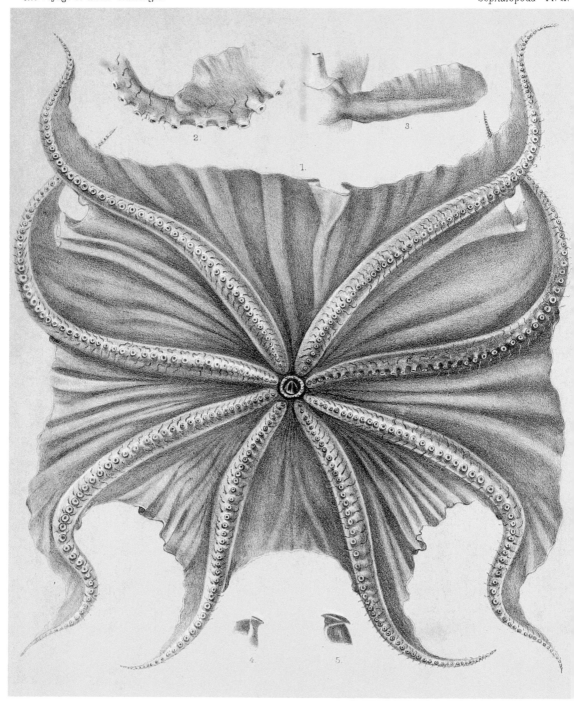

CIRROTEUTHIS PACIFICA, n. sp.

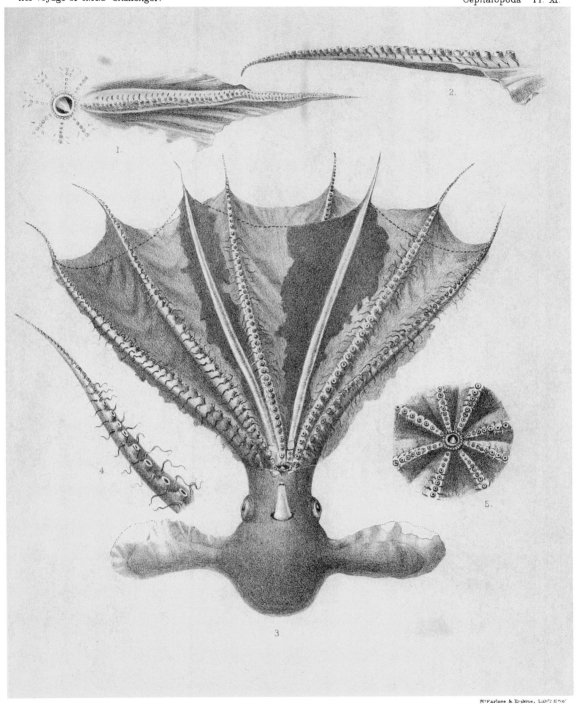

1, 2. CIRROTEUTHIS MEANGENSIS, n. sp.   3–5. STAUROTEUTHIS ?

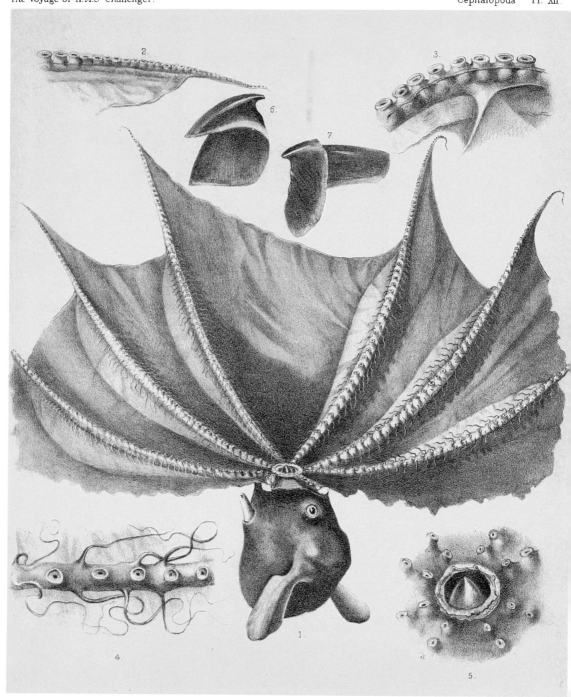

CIRROTEUTHIS MAGNA, n. sp.

第十八卷动物志十七彩图

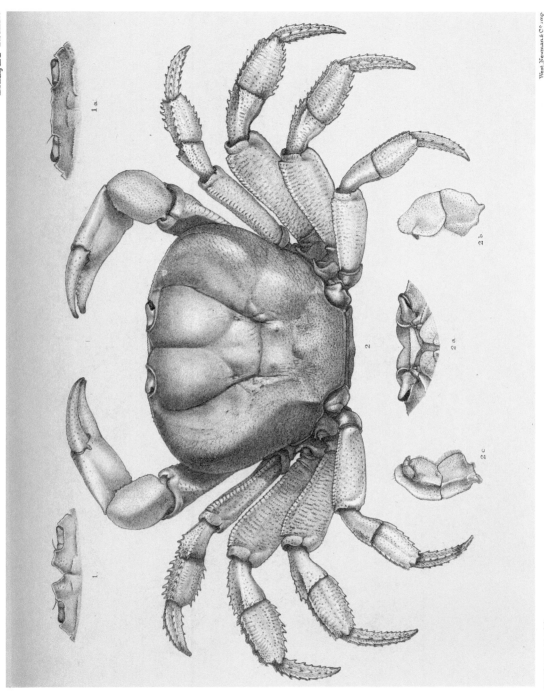

THELPHUSA – GEOCARCINUS.

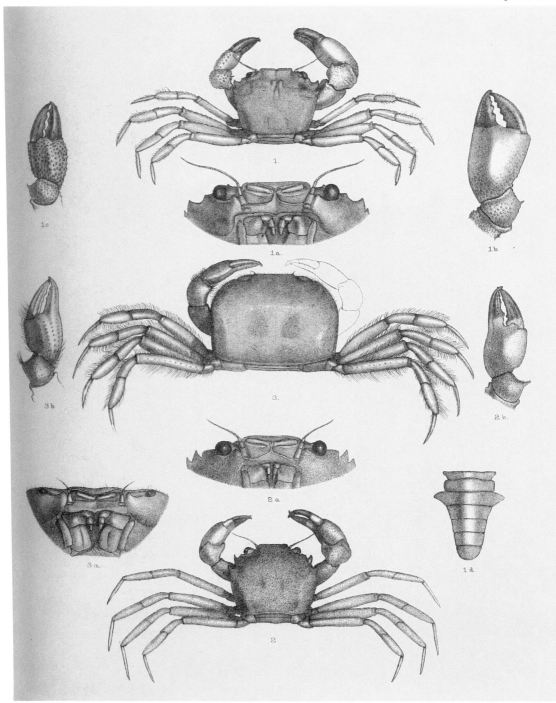

PILUMNOPLAX — CERATOPLAX.

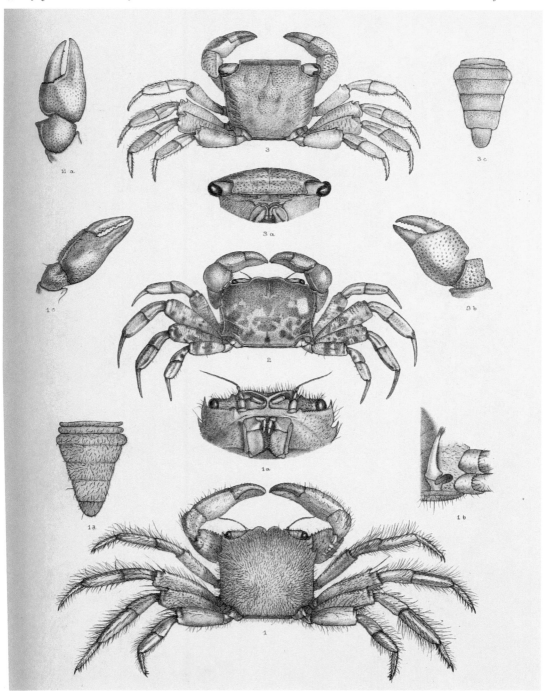

LITOCHEIRA—HELICE?—SESARMA.

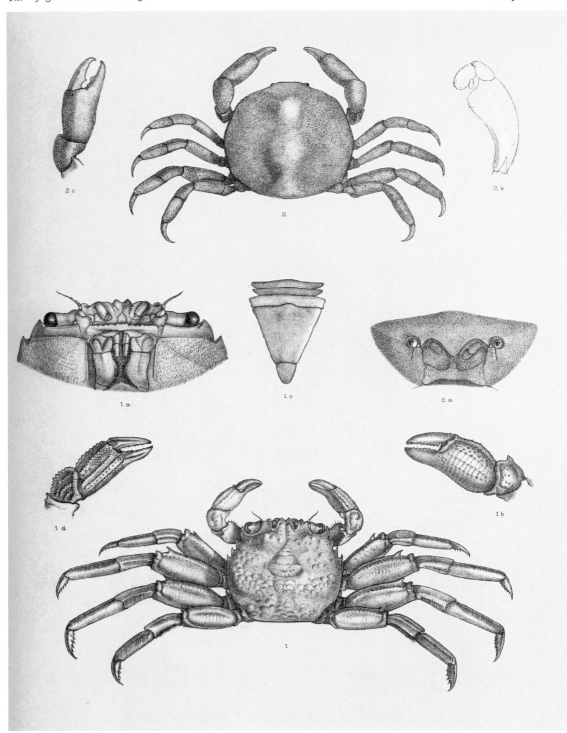

PLAGUSIA — PINNOTHERES.

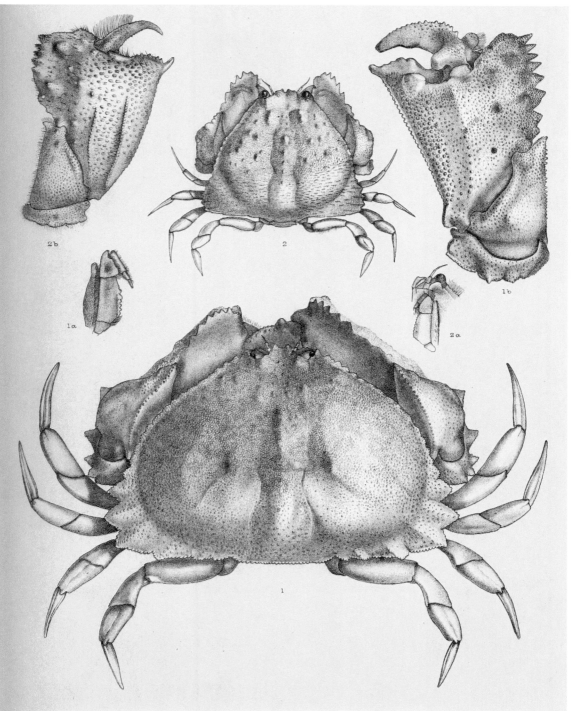

CALAPPA.

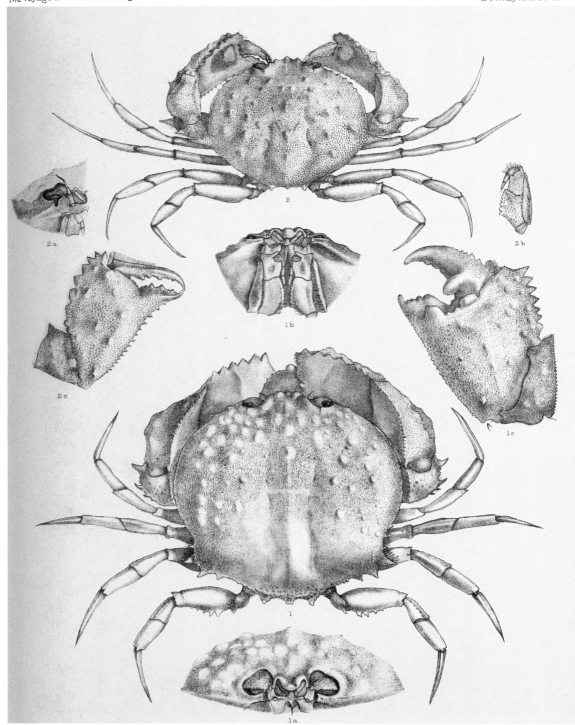

PARACYCLÖIS-MURSIA.

第二十一卷动物志十八(图版)彩图

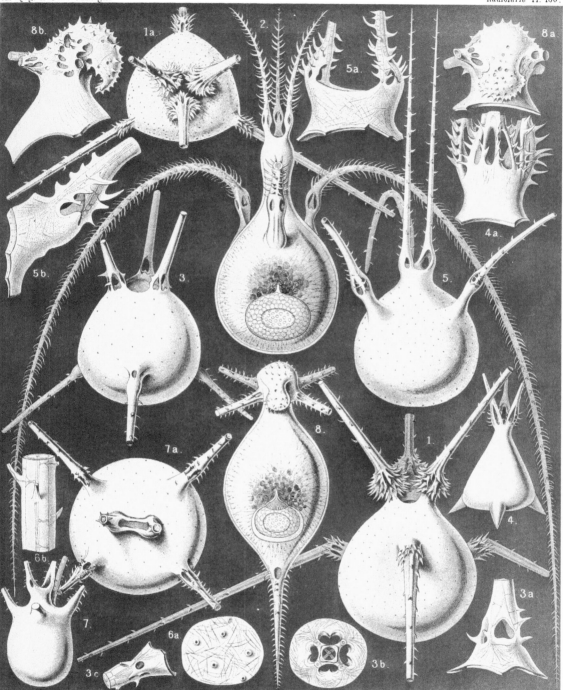

TUSCARORA.

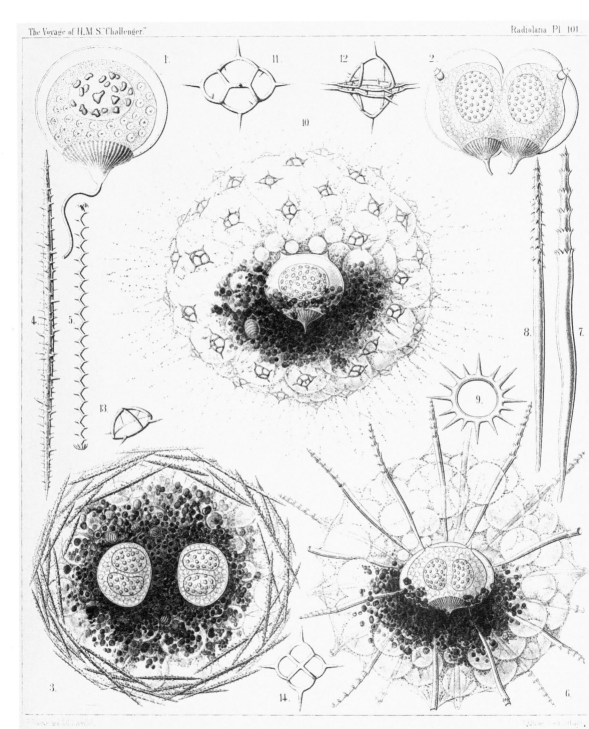

1-2. PHAEODINA, 3-5. CANNORRHAPHIS, 6-8. AULACTINIUM, 9. MESOCENA, 10-14. DICTYOCHA.

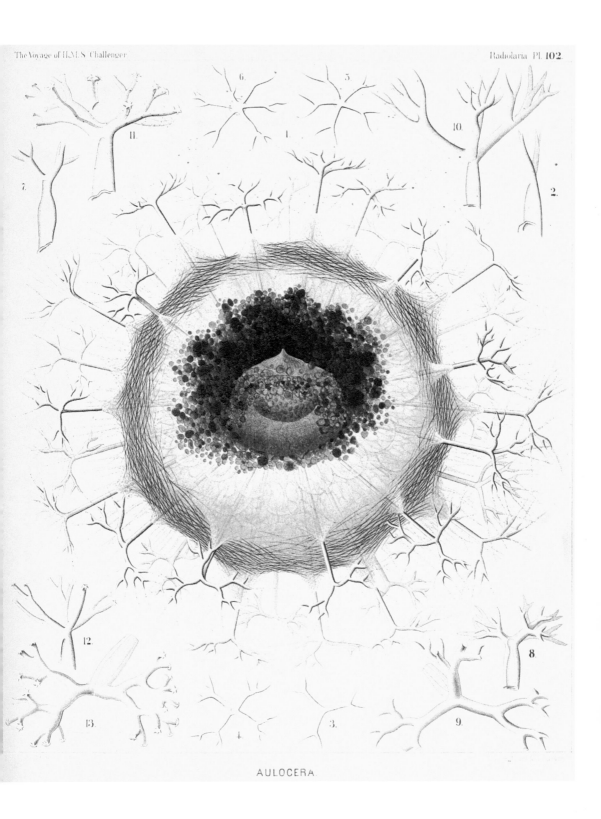

AULOCERA.

AULOGRAPHIS.

第二十三卷动物志二十彩图

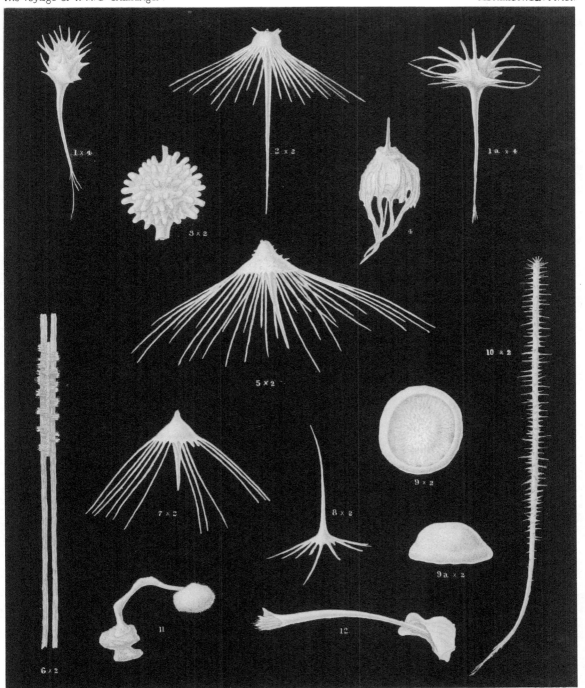

CLADORRHIZA, AXONIDERMA, CHONDROCLADIA, MELIIDERMA.

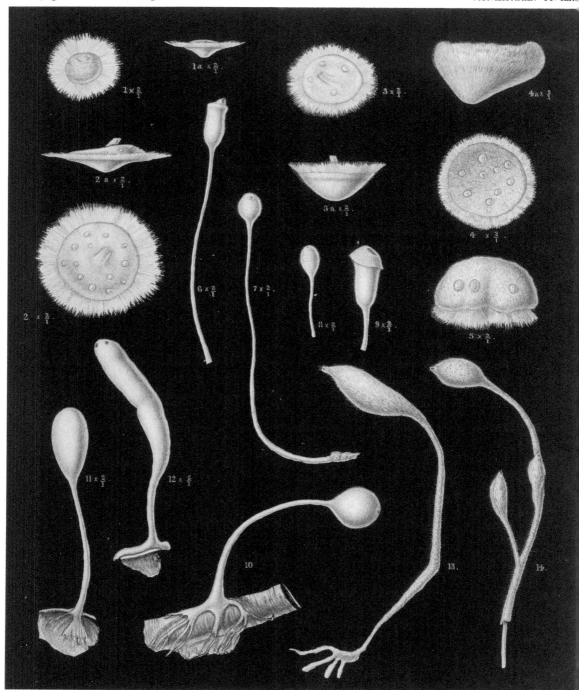

SUBERITES, TRICHOSTEMMA, STYLOCORDYLA.

第二十五卷动物志二十一(图版)彩图

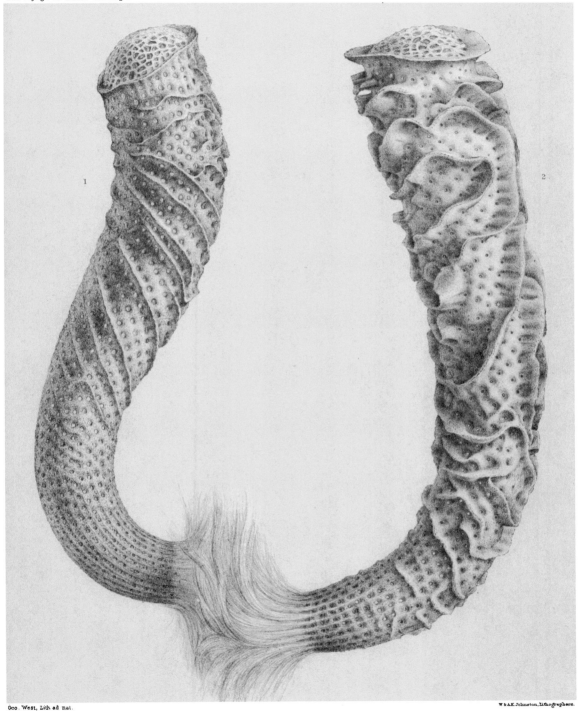

EUPLECTELLA ASPERGILLUM OWEN.

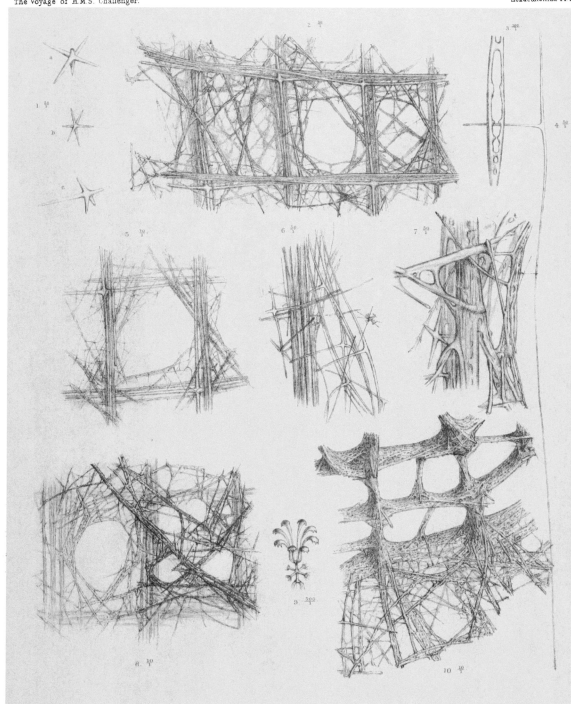

EUPLECTELLA ASPERGILLUM. OWEN.

第二十六卷动物志二十二彩图

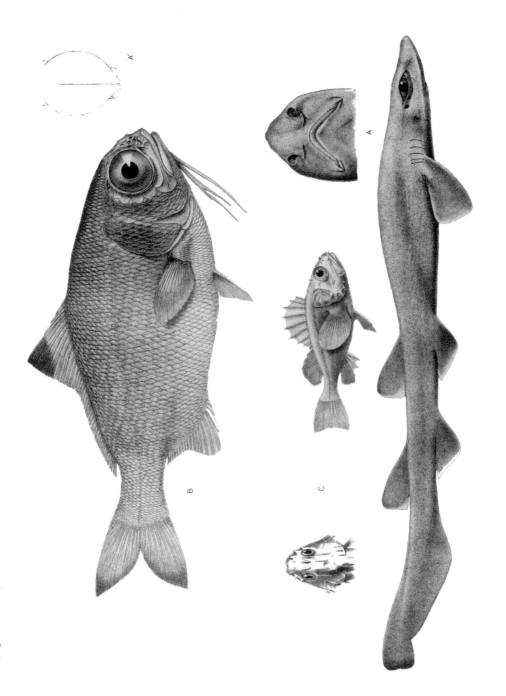

A. SCYLLIUM CANESCENS. (Faths 400.)   B. POLYMIXIA NOBILIS. (Faths 345.)   C. SETARCHES FIDJIENSIS. (Faths 215.)

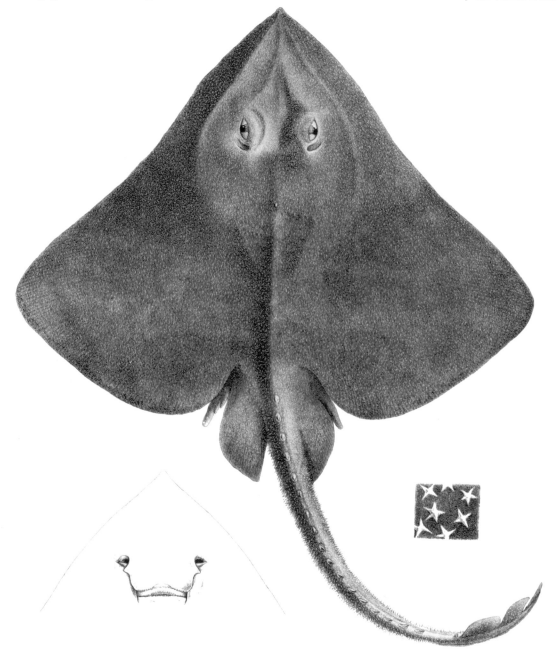

RAJA ISOTRACHYS
(Faths 565)

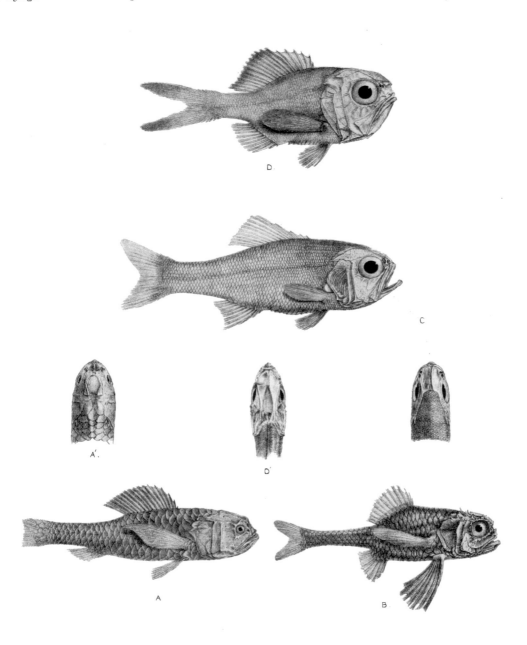

A. MELAMPHAËS TYPHLOPS.
B. MELAMPHAËS MEGALOPS.
C. TRACHICHTHYS ELONGATUS.
D. TRACHICHTHYS INTERMEDIUS.

A. BATHYDRACO ANTARCTICUS.   B. MELAMPHAËS CRASSICEPS.

A COTTUNCULUS MICROPS
(Faroe, fath. 307-608)

B COTTUNCULUS THOMSONII.
(Faroe, fath. 555)

A. CHAUNAX PICTUS.
(Faths. 215)
B. CYTTUS ABBREVIATUS.
(Faths. 400)
C. COTTUS BATHYBIUS.
(Faths. 565)

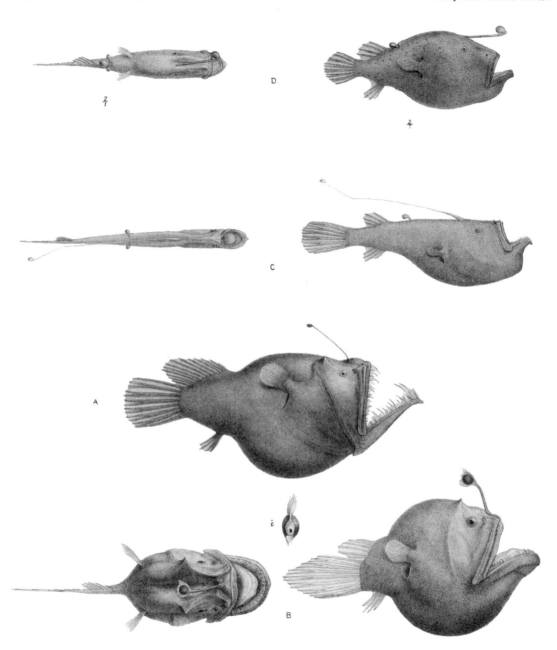

A. MELANOCETUS MURRAYI.
(Faths. 1850-2450.)
B. CERATIAS BISPINOSUS. C. CERATIAS URANOSCOPUS. D. CERATIAS CARUNCULATUS.
(Faths. 360.) (Faths. 2400.) (Faths. 345.)

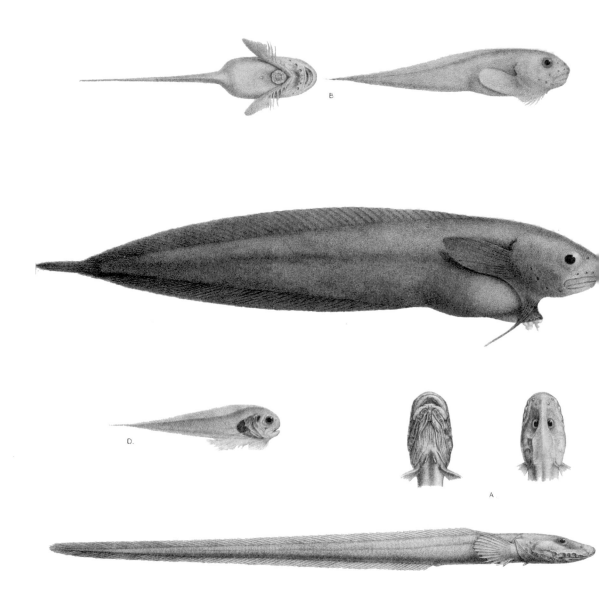

A. LYCODES MUROENA.　B. LIPARIS MICROPUS.　C. PARALIPARIS BATHYBIUS.
(Faroe Channel. Faths. 540–640.)
D. PARALIPARIS MEMBRANACEUS
(Faths. 400.)

A. LOTELLA MARGINATA.
(Faths 140)

B. MELANONUS GRACILIS
(Faths 1975)

The Voyage of H.M.S."Challenger." Deep-sea Fishes. Pl. XXIII.

A BATHYONUS TÆNIA.   B CATÆTYX MESSIERI   C DIPLACANTHOPOMA BRACHYSOMA.
(Fath. 2500)   (Fath. 345.)   (Fath. 350)

R. Mintern del et lith.   Mintern Bros. imp.

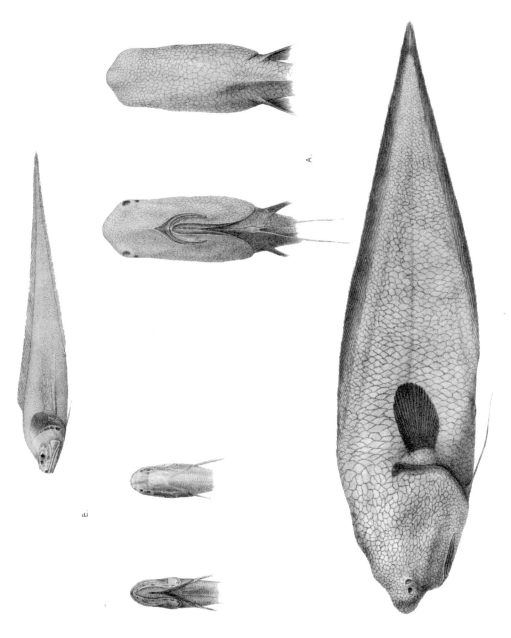

A TYPHLONUS NASUS (Fahs. 2440)
B MIXONUS LATICEPS (Fahs. 2500)

A. APHYONUS GELATINOSUS.

B. BATHYPTEROIS LONGICAUDA.

Deep-sea Fishes. Pl. XXVIII.

The Voyage of H.M.S. Challenger.

P. Smit del. et lith.

Mintern Bros. imp.

A. MACRURUS FASCIATUS.
(*Faths.* 140.)

B. MACRURUS HOLOTRACHYS.
(*Faths.* 600.)

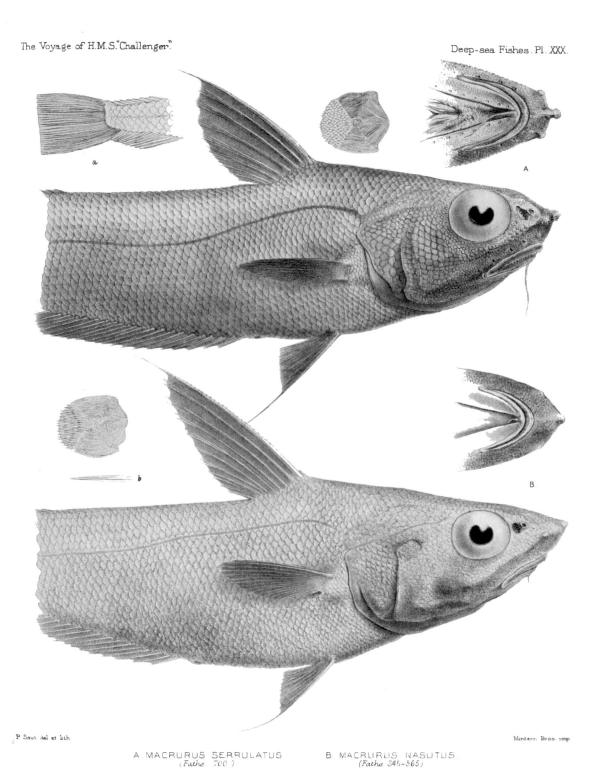

A. MACRURUS SERRULATUS
(Faths 700)

B. MACRURUS NASUTUS.
(Faths 345-565)

MACRURUS LEPTOLEPIS
(Faths 850)

A. MACRURUS SCLERORHYNCHUS. B. MACRURUS BAIRDII. C. MACRURUS ÆQUALIS.
(Faths 1090.) (Faths 160–740.) (Faths 600.)

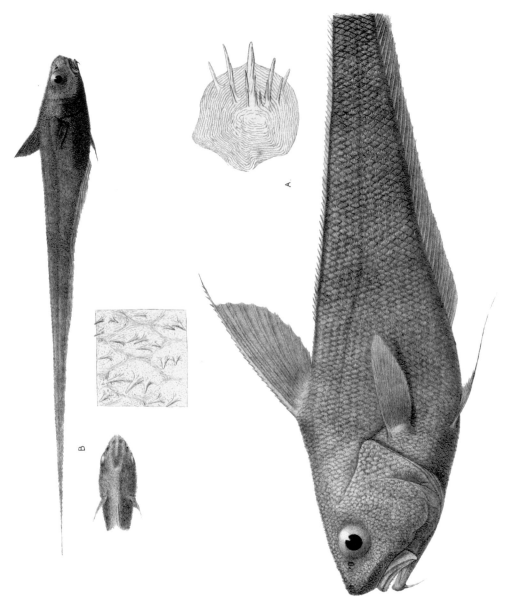

A. MACRURUS ASPER. (Faths. 1875.)
B. MACRURUS VILLOSUS. (Faths. 345–500.)

MACRURUS CRASSICEPS.
(Fibra 520.)

A. MACRURUS LIOCEPHALUS
(Faths 1875)
B. MACRURUS FERNANDEZIANUS.
(Faths 1875)

A. BATHYGADUS COTTOIDES. B. BATHYGADUS MULTIFILIS. C. LYCONUS PINNATUS.
(Faths. 700.) (Faths. 500.)
D. ONUS CARPENTERI.
(Faths. 180.)

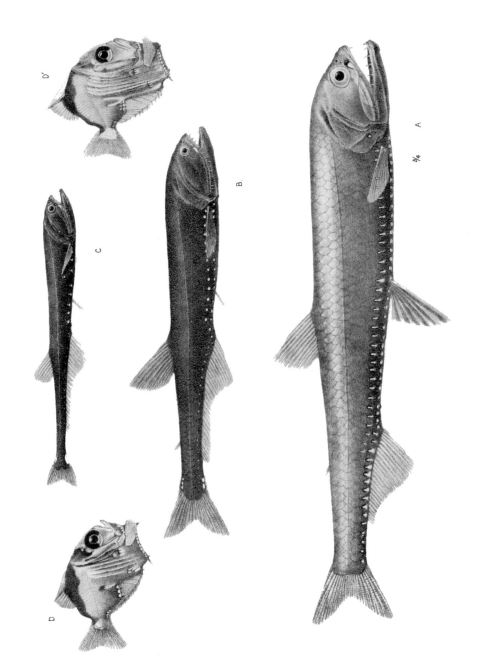

A. PHOTICHTHYS ARGENTEUS. B. GONOSTOMA ELONGATUM (Faths. 800.) C. GONOSTOMA GRACILE (Faths. 365.)
D, D'. STERNOPTYX DIAPHANA.

Deep-sea Fishes Pl. XLVIII.

The Voyage of H.M.S."Challenger."

A BATHYPTEROIS LONGIPES.
(Fathes 2650.)

B BATHYPTEROIS LONGIFILIS.

R. Mintern del et lith.

Mintern Bros. imp.

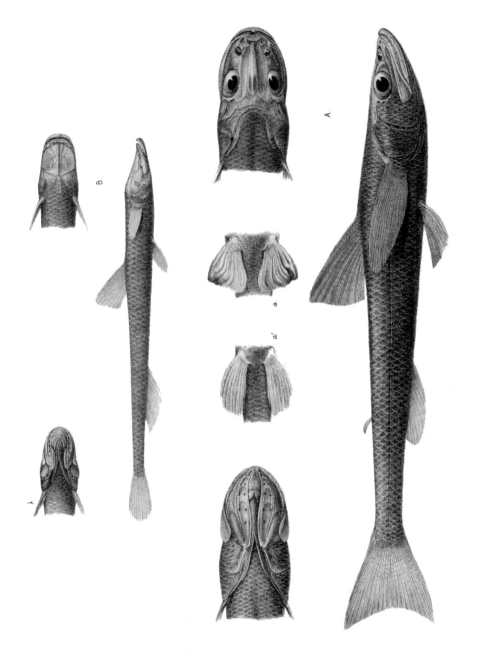

A. CHLOROPHTHALMUS GRACILIS. B. IPNOPS MURRAYI.

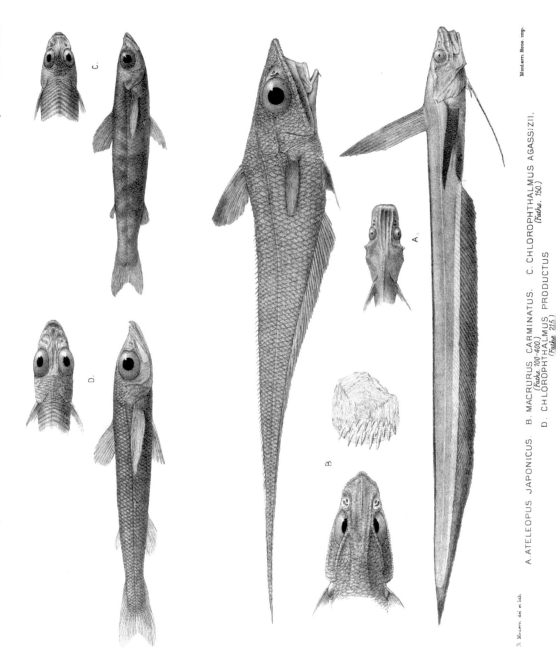

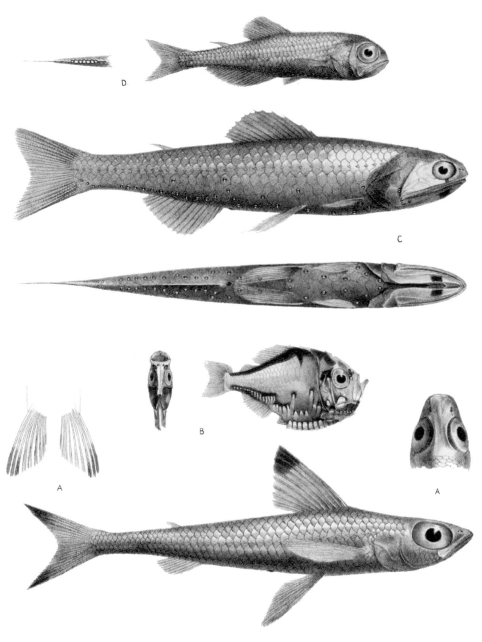

A. CHLOROPHTHALMUS NIGRIPINNIS. B. POLYIPNUS SPINOSUS
(Faths. 120.) (Faths. 255.)
C. SCOPELUS ENGRAULIS. D. SCOPELUS ANTARCTICUS
(Faths. 255.) (Faths. 1975.?)

A. ODONTOSTOMUS HYALINUS.   B. NANNOBRACHIUM NIGRUM.   C, C'. OMOSUDIS LOWII.
(Faths. 500.)   (Faths. 500.)
D. IDIACANTHUS FEROX.
(Faths. 2750.)

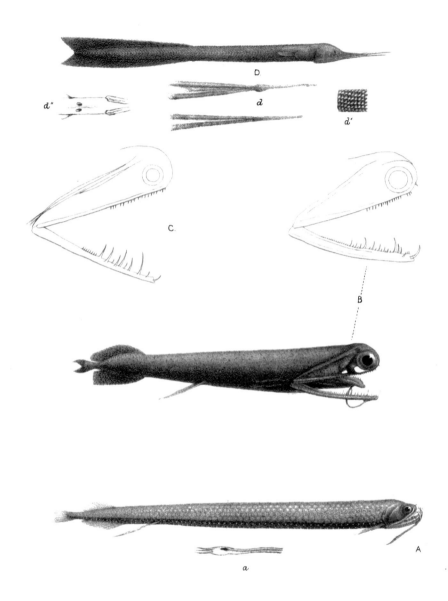

A STOMIAS AFFINIS (Faths. 450.)
B. MALACOSTEUS INDICUS (Faths. 500.)
C. MALACOSTEUS NIGER
D. CYEMA ATRUM (Faths. 1500–1800.)

A TRACHICHTHYS TRAILLII   B ARGENTINA ELONGATA

A. BATHYTROCTES MICROLEPIS (Faths 1090) B. BATHYTROCTES MACROLEPIS (Faths 2750)

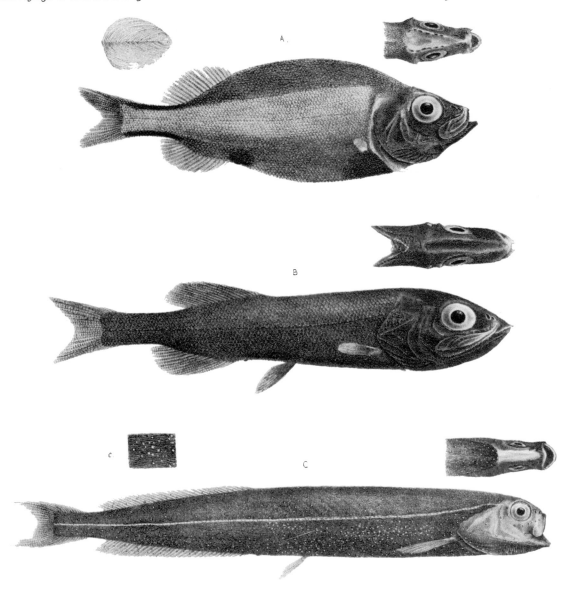

A. PLATYTROCTES APUS
(Faths 1500)
B. BATHYTROCTES ROSTRATUS
(Faths 675)
C. XENODERMICHTHYS NODULOSUS
Faths 345

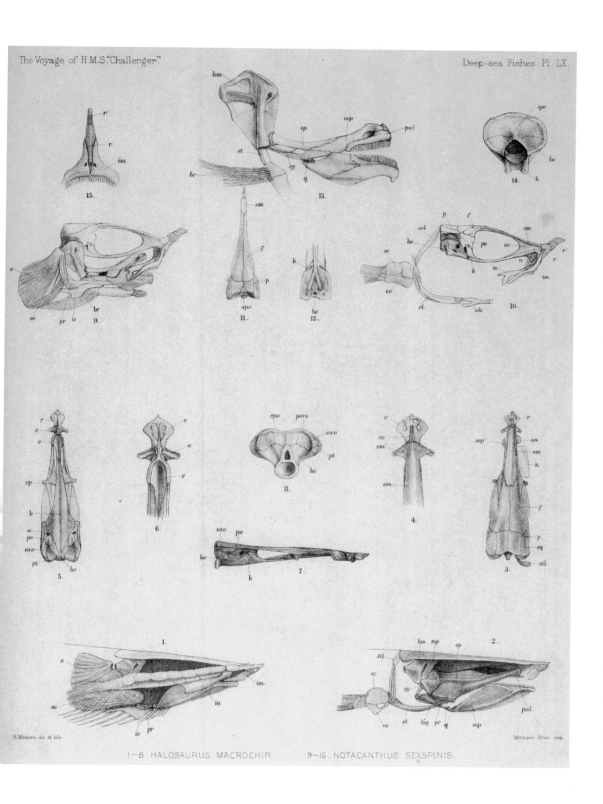

1–8. HALOSAURUS MACROCHIR.   9–15. NOTACANTHUS SEXSPINIS.

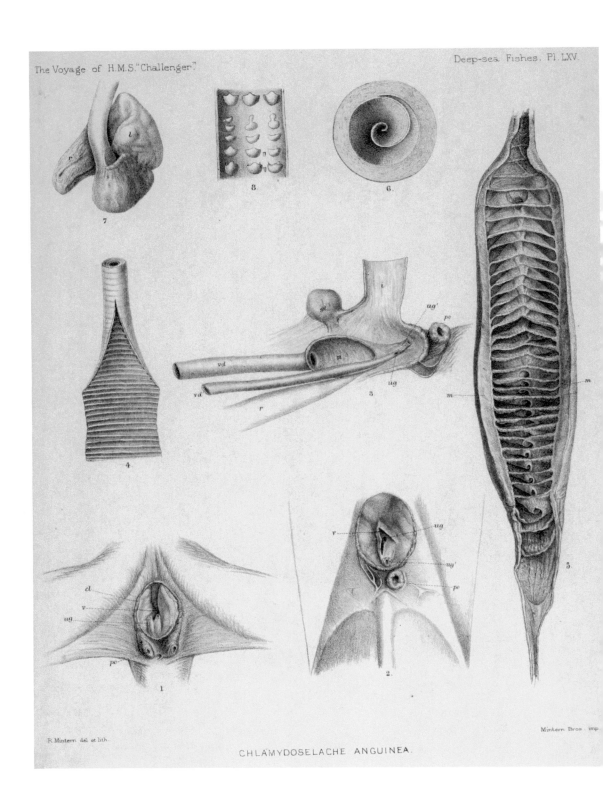

第三十一卷动物志二十六(图版)彩图

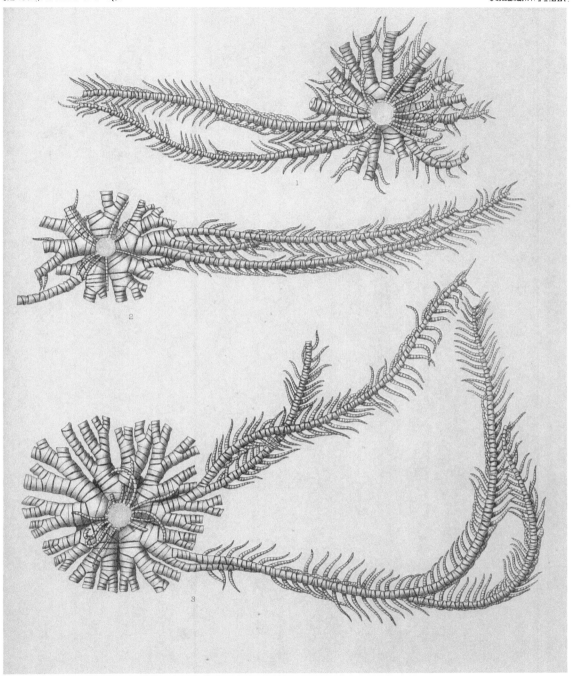

1. ACTINOMETRA SIMPLEX, sp. n.
2. ACTINOMETRA ROTALARIA Lam., sp. 3. ACTINOMETRA VALIDA, sp. n.

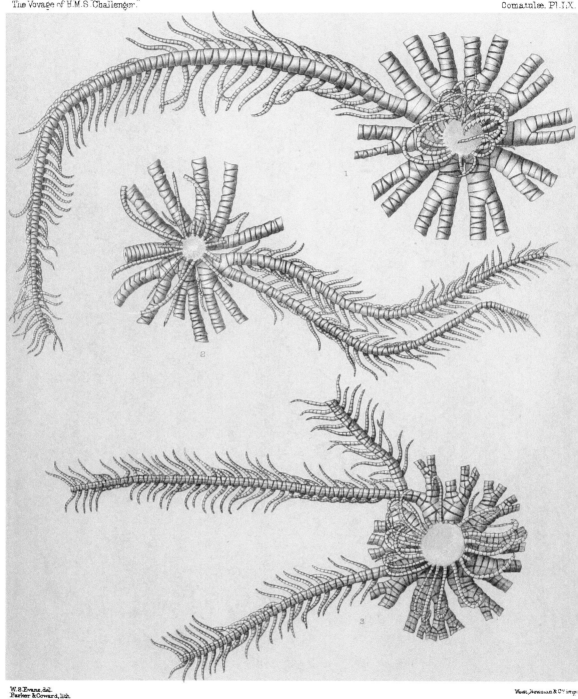

1, 2. ACTINOMETRA COPPINGERI, Bell.   3. ACTINOMETRA LINEATA, sp. n.

第三十三卷动物志二十八彩图

ATHORYBIA OCELLATA

ATHORYBIA OCELLATA

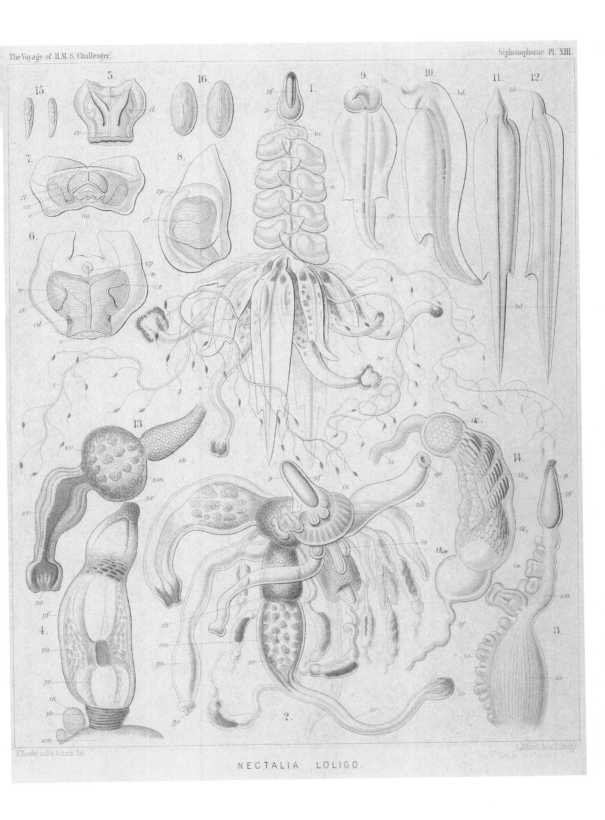

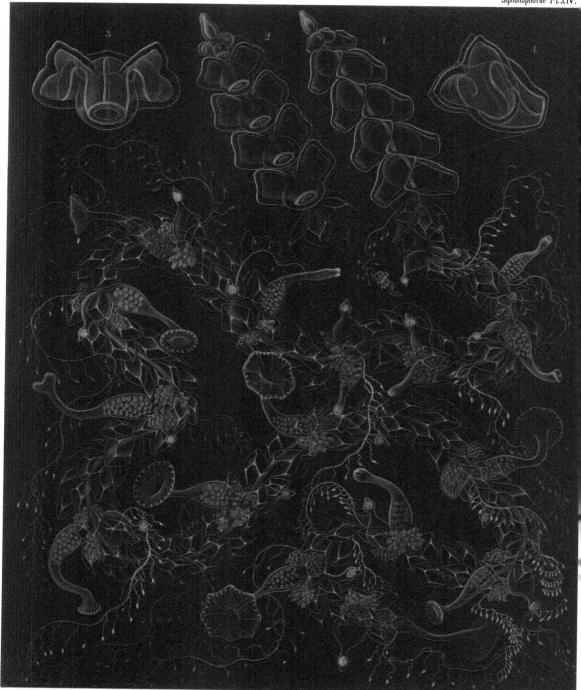

ANTHEMODES ORDINATA.

1–5. CYSTALIA MONOGASTRICA. 6–8. EPIBULIA RITTERIANA.

NECTOPHYSA WYVILLEI.

CANNOPHYSA MURRAYANA.

SALACIA POLYGASTRICA.

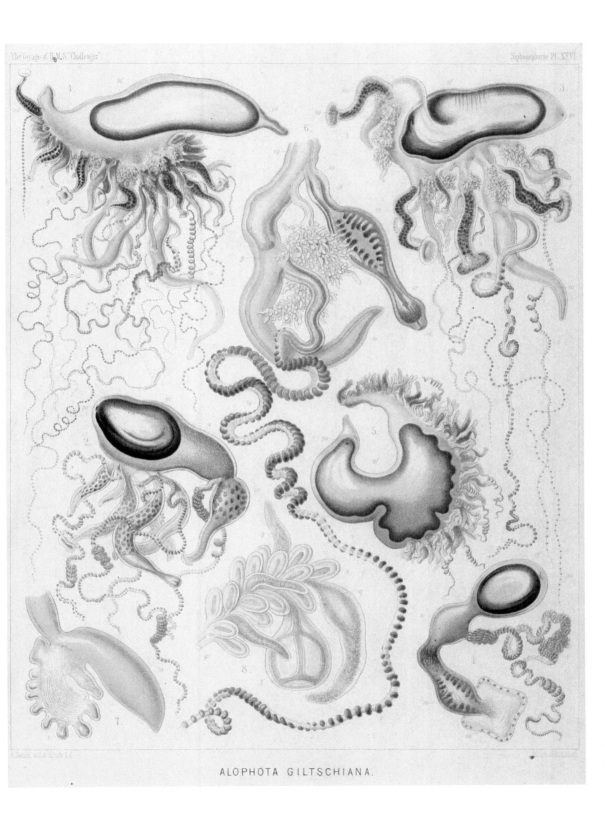

ALOPHOTA GILTSCHIANA.

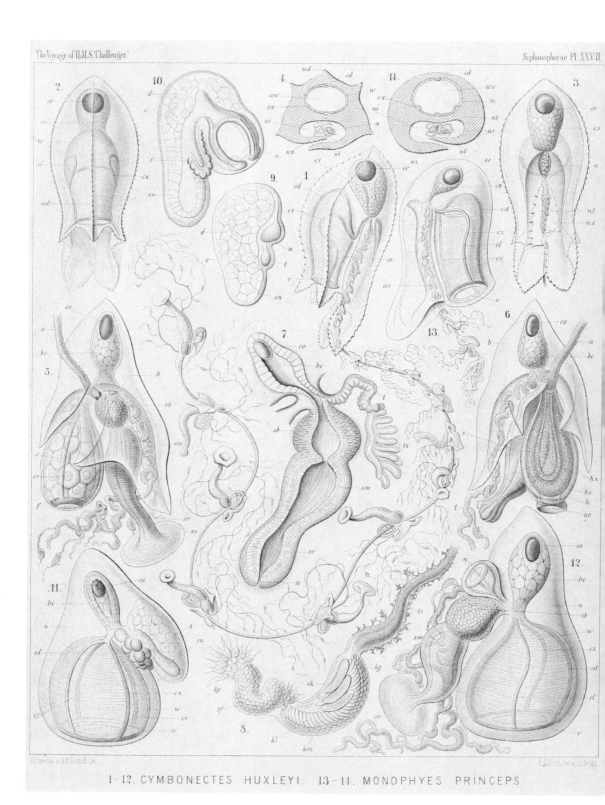

MITROPHYES PELTIFERA.

1–8. POLYPHYES UNGULATA. 9–14. VOGTIA KÖLLIKERI.

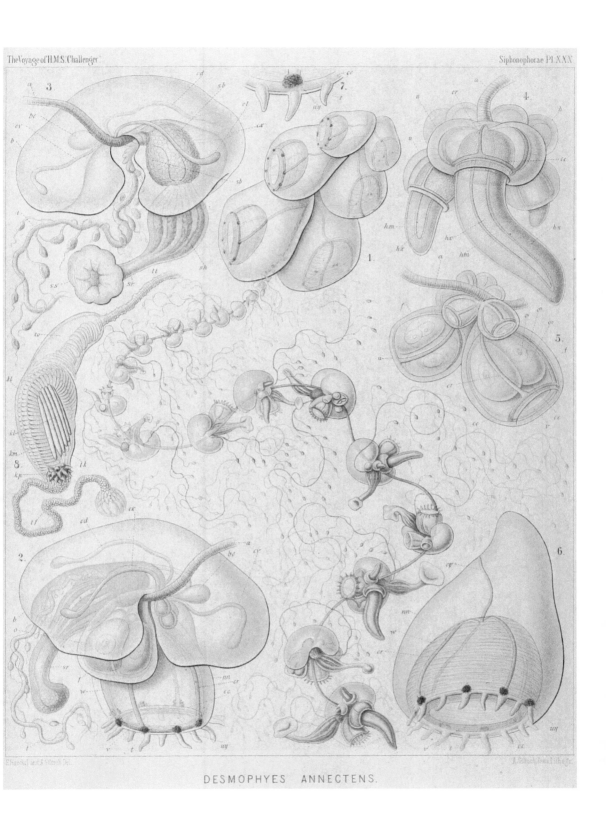

DESMOPHYES ANNECTENS.

PRAYA GALEA.

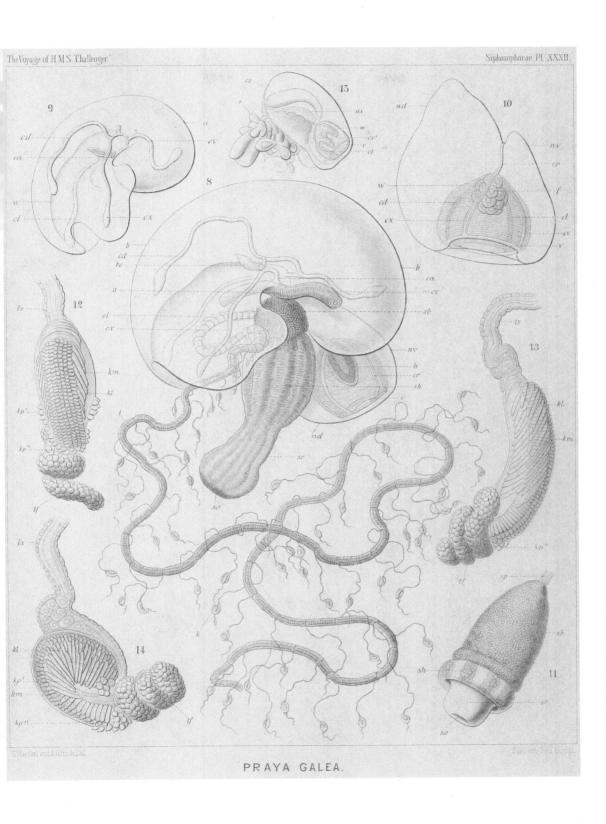

PRAYA GALEA.

DISCONALIA GASTROBLASTA

第三十七卷动物志三十(图版)彩图

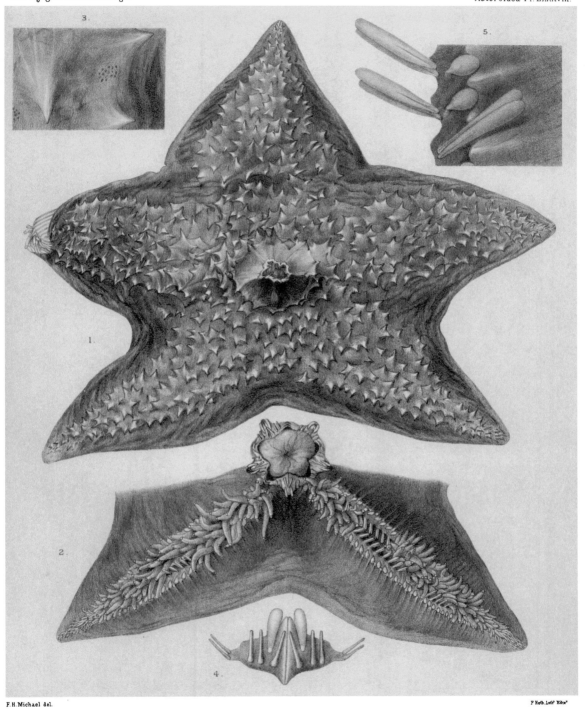

HYMENASTER CARNOSUS, n.sp.

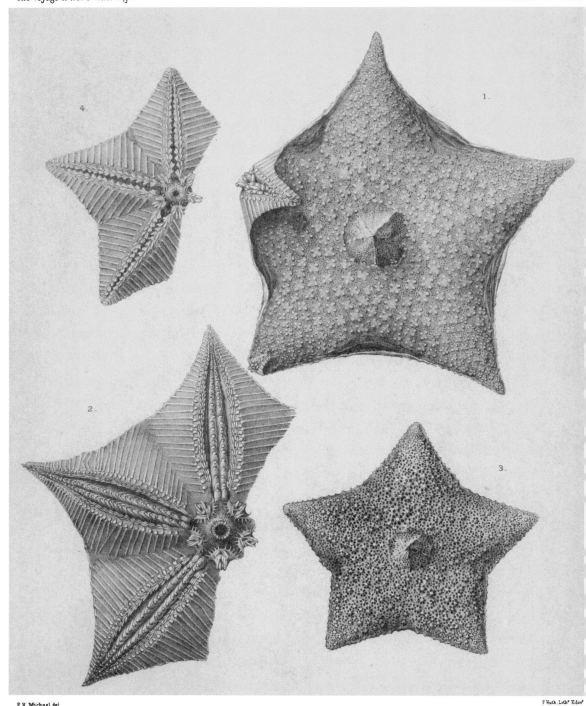

1-2. HYMENASTER CRUCIFER, n.sp.   3-4. HYMENASTER ANOMALUS, n.sp.

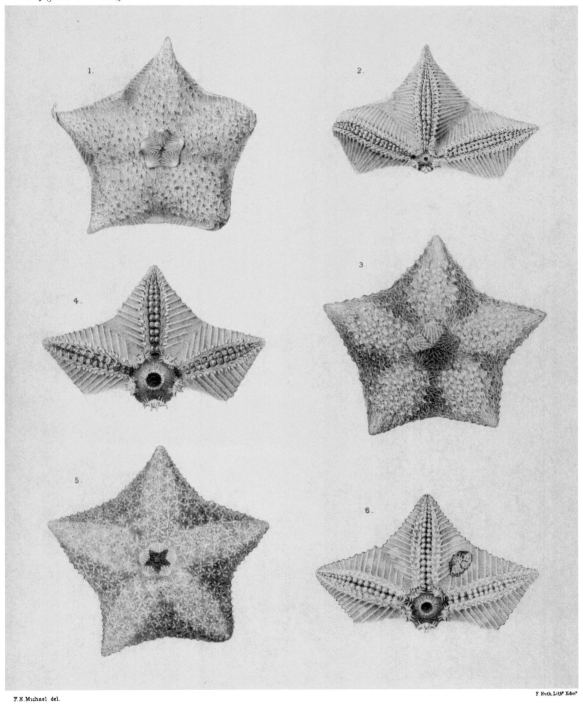

1-2. HYMENASTER GRANIFERUS, n.sp. 3-4. HYMENASTER COCCINATUS, n.sp.
5-6. HYMENASTER PRÆCOQUIS, n.sp.

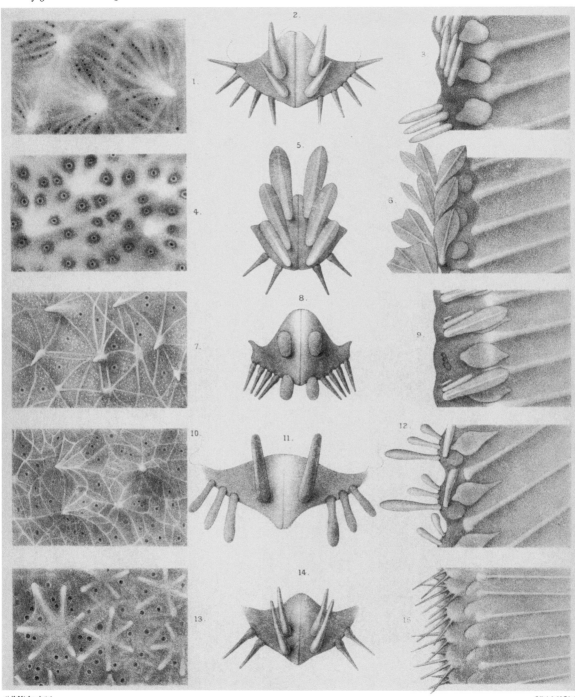

1-3. HYMENASTER CRUCIFER, n.sp.   4-6. HYMENASTER ANOMALUS, n.sp.   7-9. HYMENASTER GRANIFERUS, n.sp.
10-12. HYMENASTER COCCINATUS, n.sp.   13-15. HYMENASTER PRÆCOQUIS, n.sp.

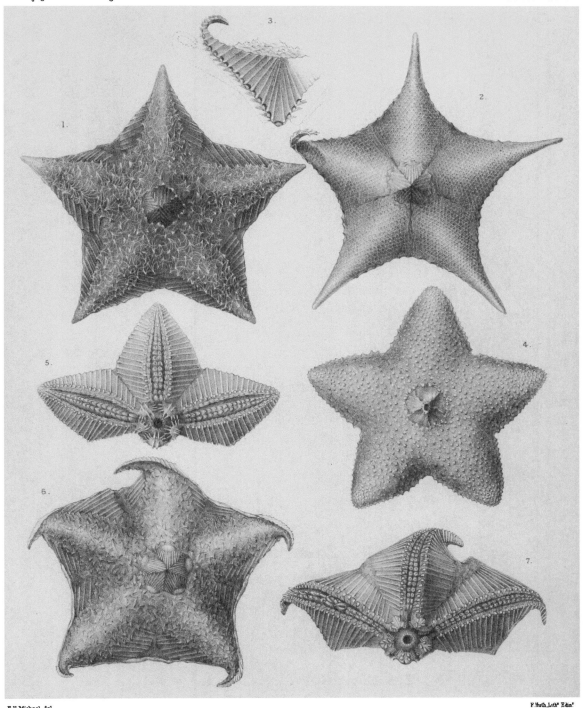

1. HYMENASTER PULLATUS, n.sp.   2-3. HYMENASTER GEOMETRICUS, n.sp.
4-5. HYMENASTER LATEBROSUS, n.sp.   6-7. HYMENASTER MEMBRANACEUS, Wyville Thomson.

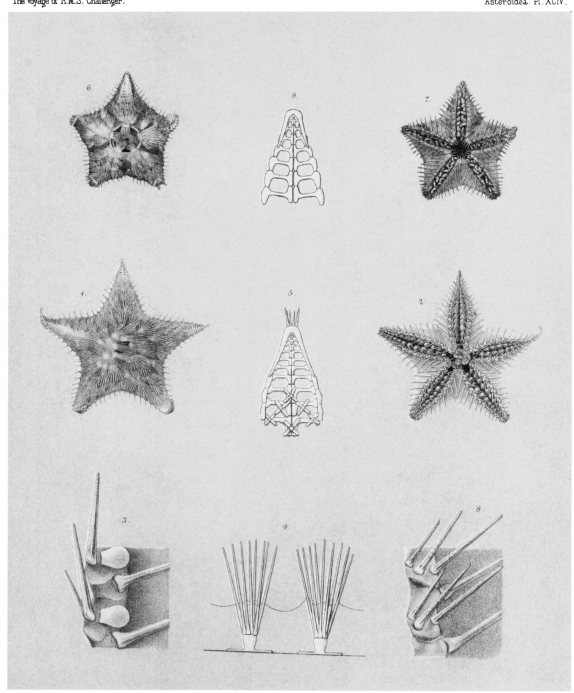

1–5. BENTHASTER WYVILLE-THOMSONI, n.sp.   6–9. BENTHASTER PENICILLATUS, n.sp.

第四十卷动物志三十二彩图

SIPHONOGORGIA.

SIPHONOGORGIA.

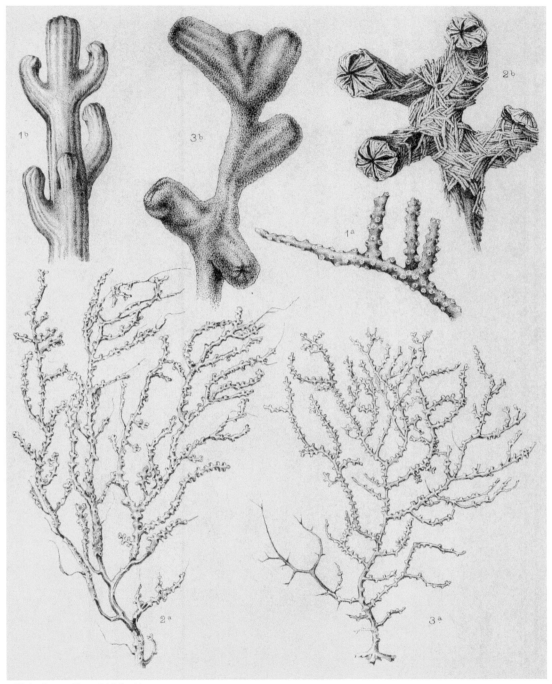

CARIJOA. ANTHOGORGIA. BEBRYCE.

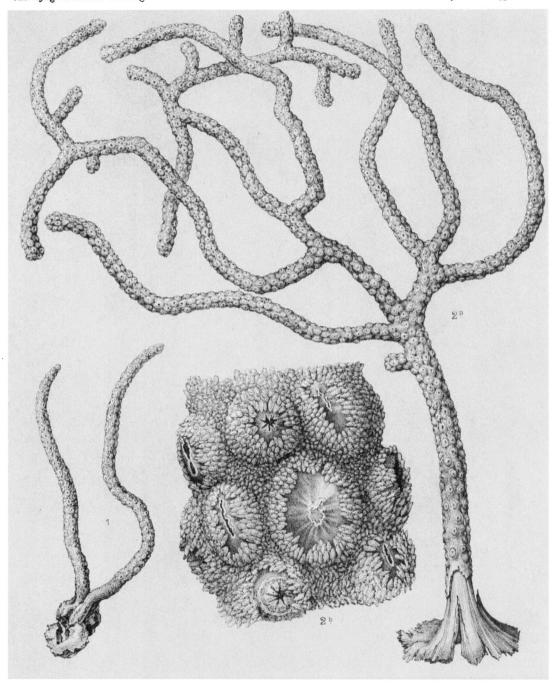

ECHINOGORGIA. EUNICEA.

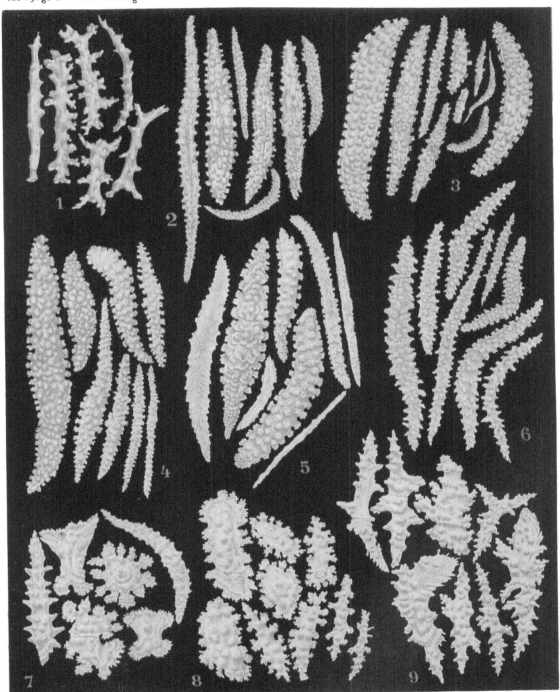

TELESTO (CARIJOA). SIPHONOGORGIA. ANTHOGORGIA. BEBRYCE. ECHINOGORGIA. EUNICEA.

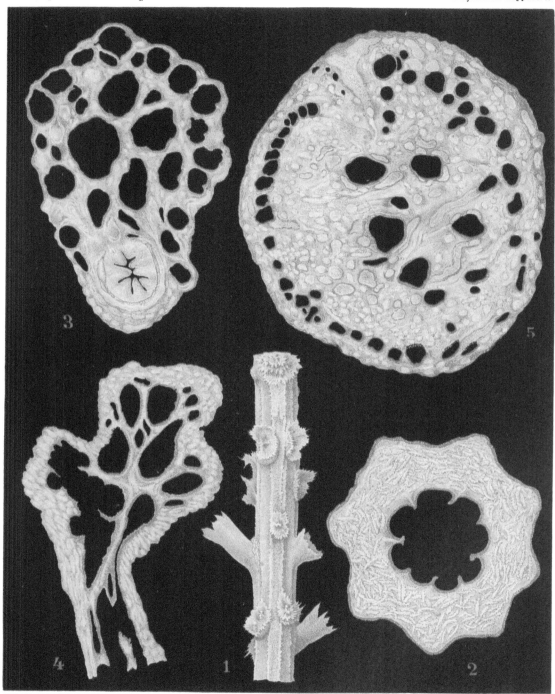

TELESTO (CARIJOA). SIPHONOGORGIA.

1-5. STANNOPHYLLUM, 1. S. ZONARIUM, 2. S. RADIOLARIUM, 3. S. PERTUSUM, 4. S. VENOSUM, 5. S. GLOBIGERINUM.

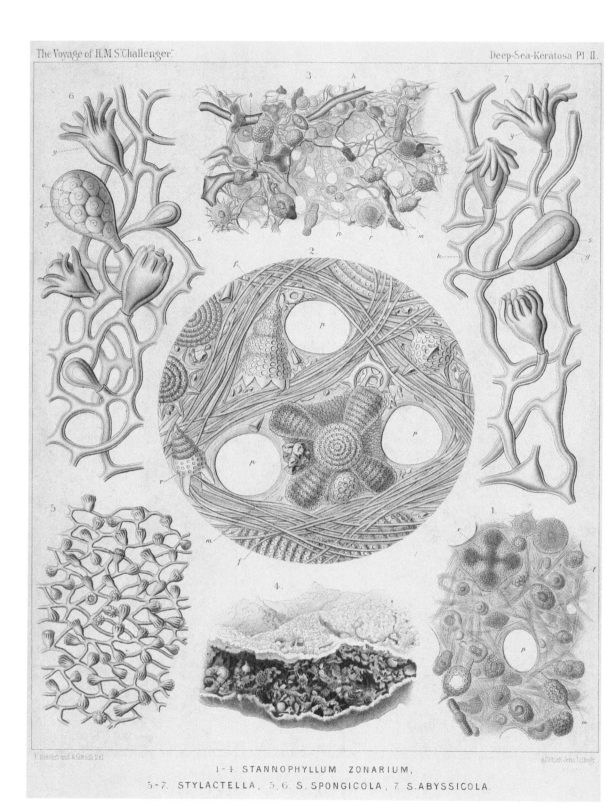

1–4. STANNOPHYLLUM ZONARIUM,
5–7. STYLACTELLA, 5,6. S. SPONGICOLA, 7. S. ABYSSICOLA.

1–8. PSAMMOPHYLLUM 1–4. P. ANNECTENS. 5–8. P. FLUSTRACEUM.
9. HALISIPHONIA SPONGICOLA.

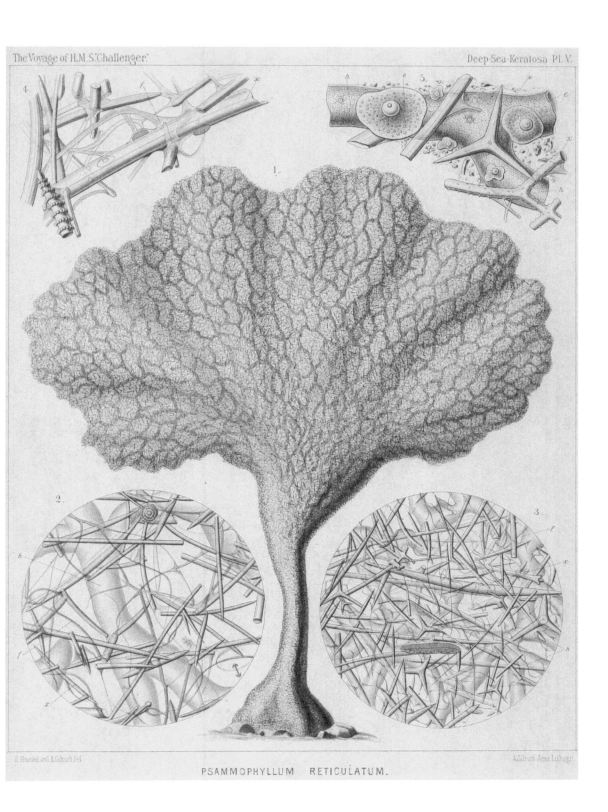

PSAMMOPHYLLUM RETICULATUM.

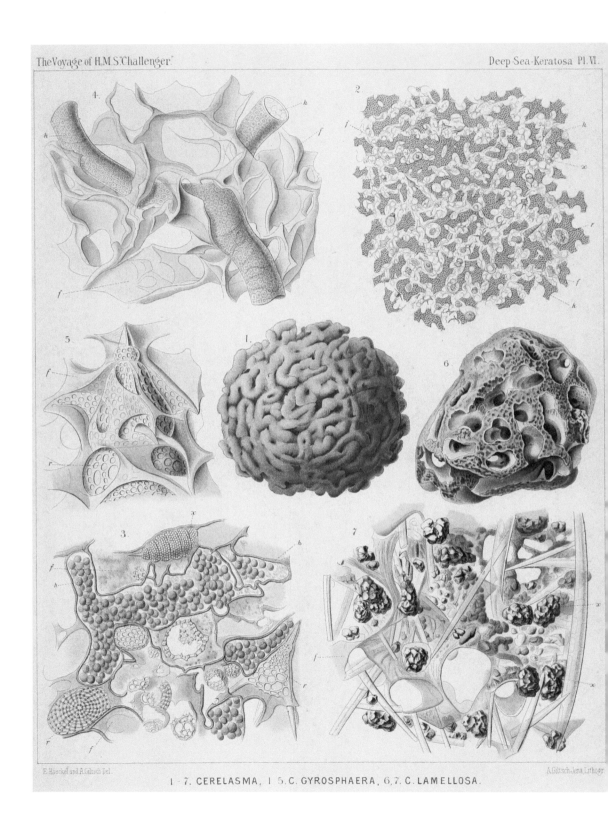

1–7. CERELASMA, 1 5. C. GYROSPHAERA, 6,7. C. LAMELLOSA.

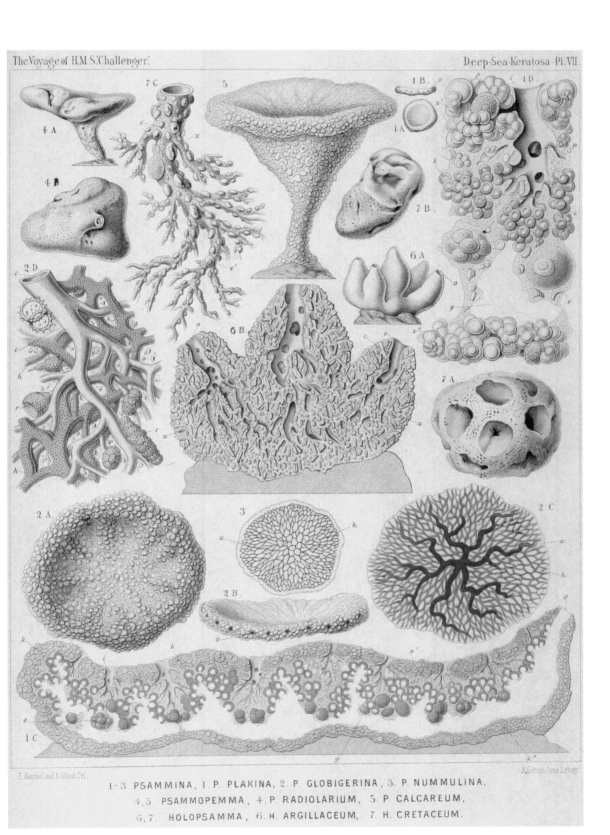

1-3. PSAMMINA, 1. P. PLAKINA, 2. P. GLOBIGERINA, 3. P. NUMMULINA,
4,5. PSAMMOPEMMA, 4. P. RADIOLARIUM, 5. P. CALCAREUM,
6,7. HOLOPSAMMA, 6. H. ARGILLACEUM, 7. H. CRETACEUM.

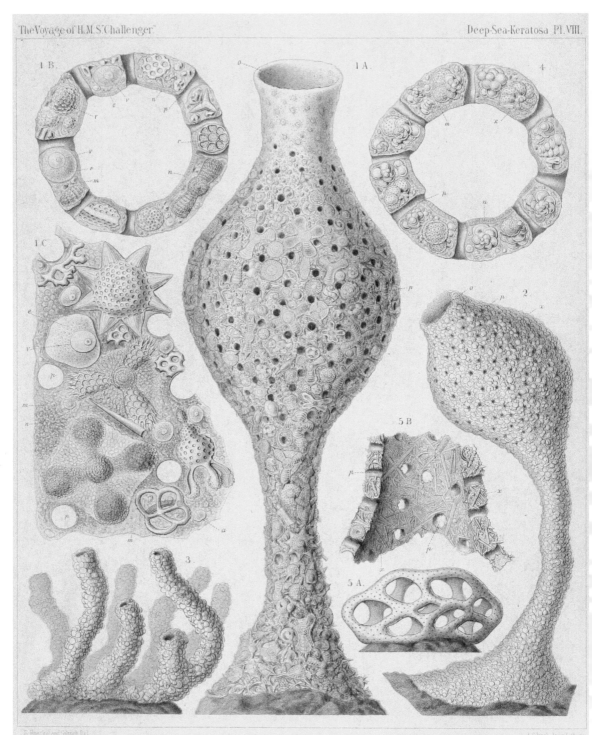

第四十二卷植物志二彩图

The Voyage of H.M.S. Challenger — *Diatomaceæ* Pl. I.

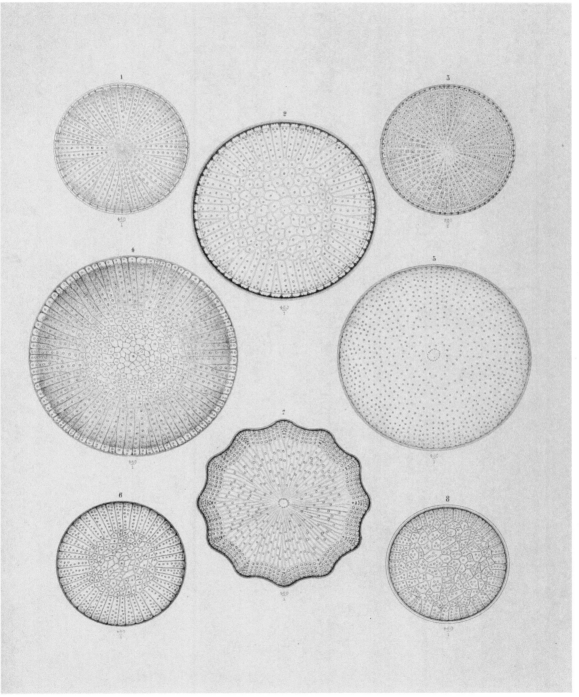

Cesare Cerù del.   Castracane direxit   Micheletti inc.

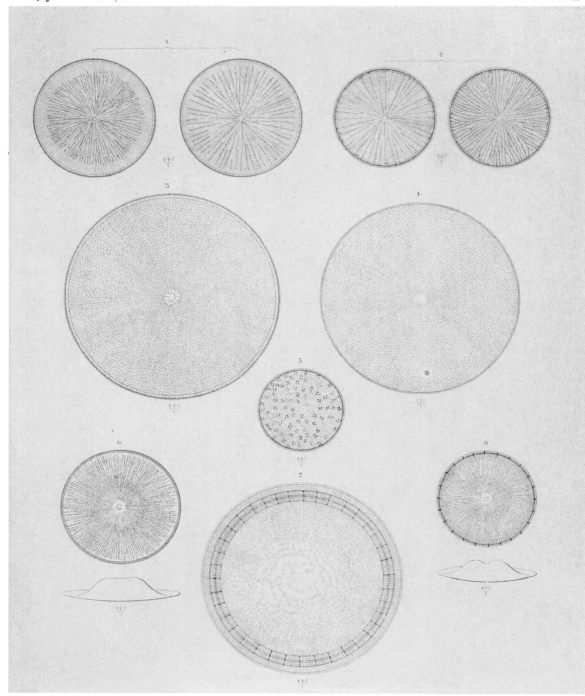

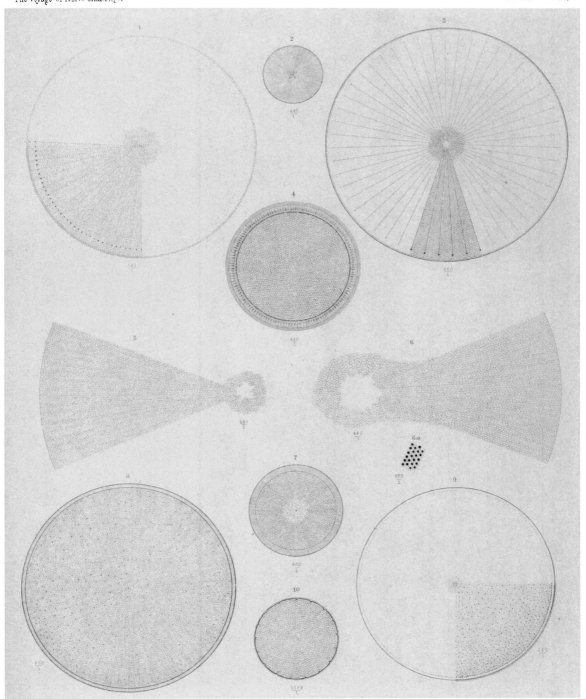

The Voyage of H.M.S. Challenger. *Diatomaceæ* Pl. V

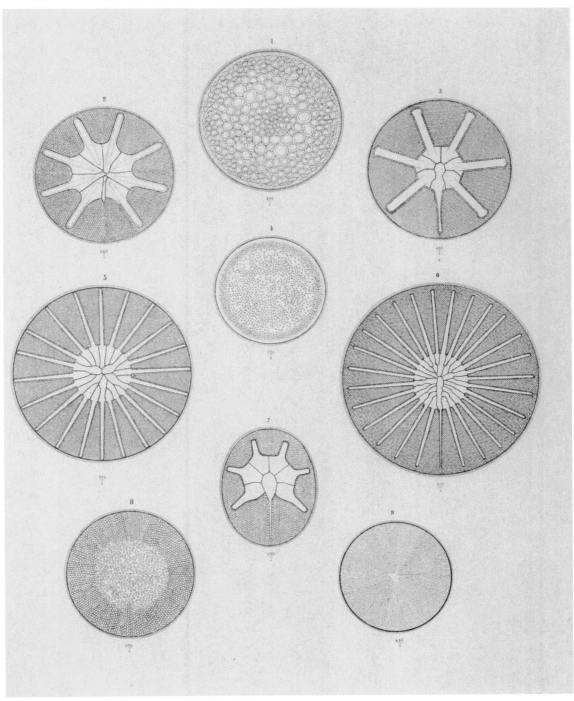

Cesare Ceri ad. Castracane direxit Micheletti inc.

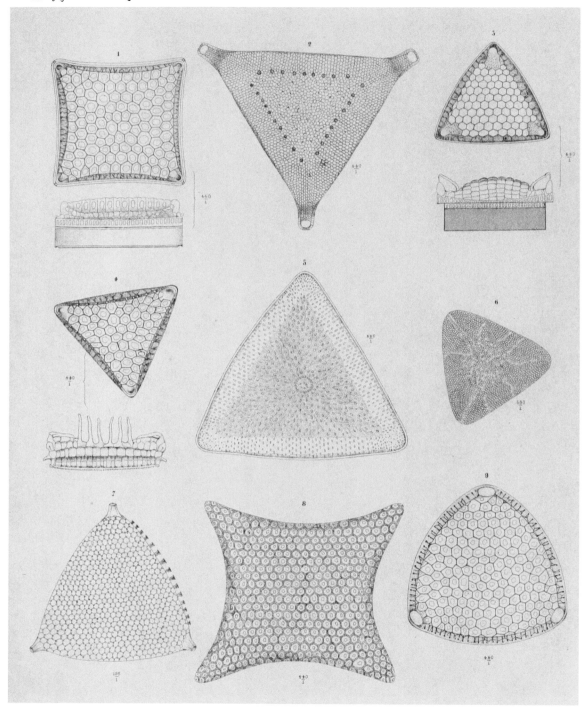

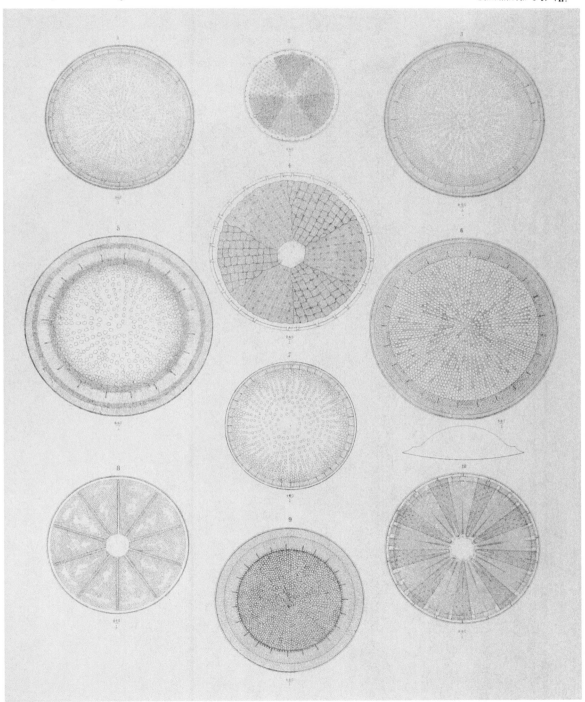

The Voyage of H.M.S. Challenger.

Diatomaceæ. Pl. VIII

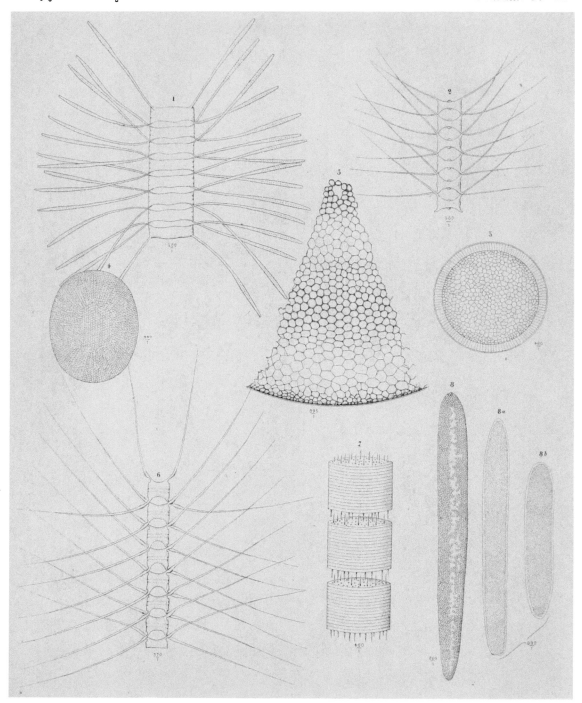

Cesare Cerù del.

Castracane diresse.

Micheletti inc.

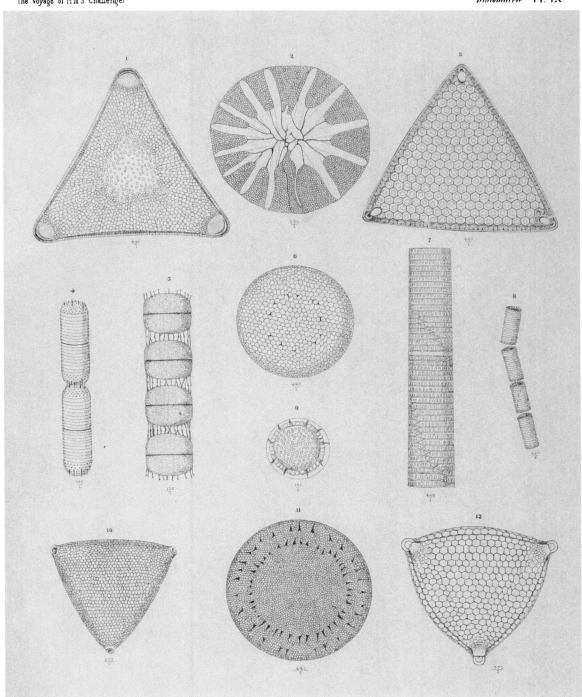

The Voyage of H.M.S. Challenger. *Diatomaceæ* Pl. X

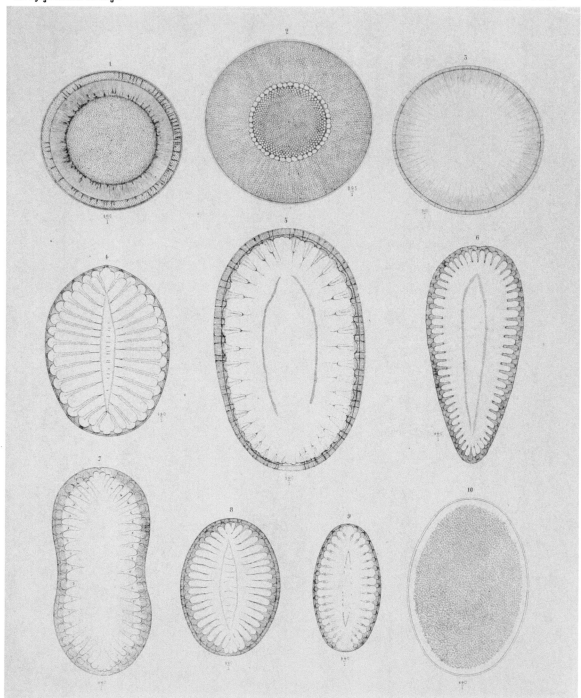

Castracane diresse   Cesare Cerù del.   Micheletti inc.

The Voyage of H.M.S."Challenger" Diatomaceæ Pl. XI

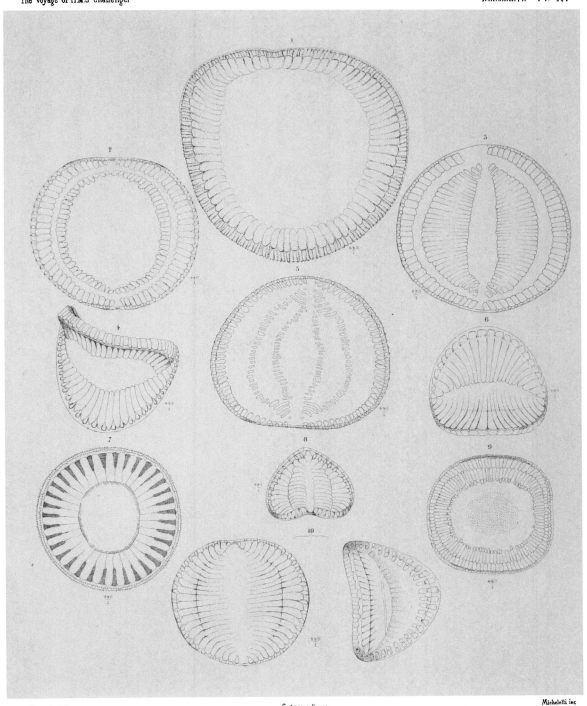

Cesare Cerù del. Castracane diresse Micheletti inc

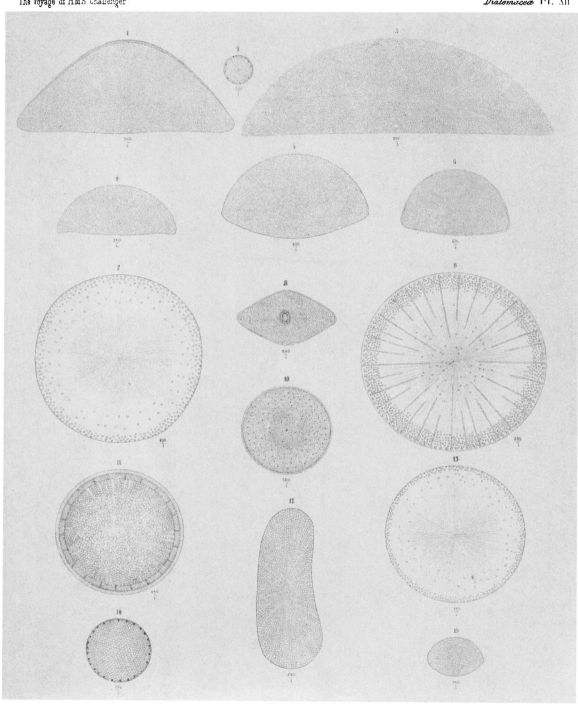

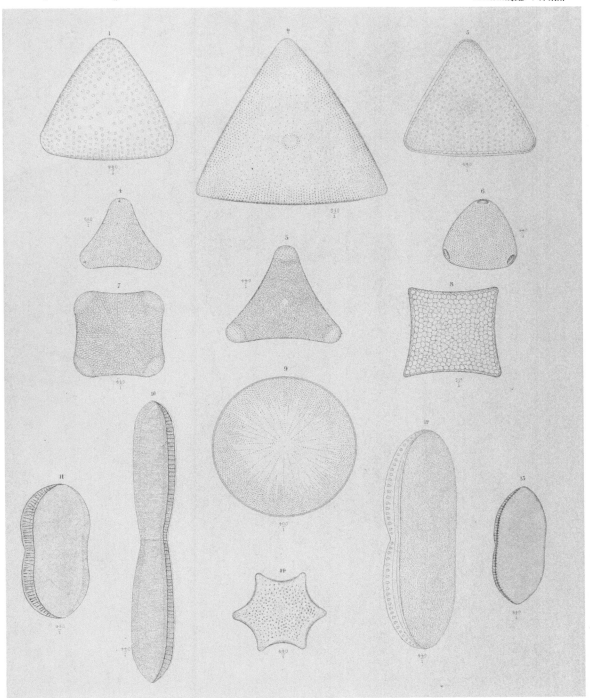

The Voyage of H.M.S. Challenger   Diatomaceæ Pl. XIV.

Cesare Cavi ad.   Castracane diresse.   Micheletti inc.

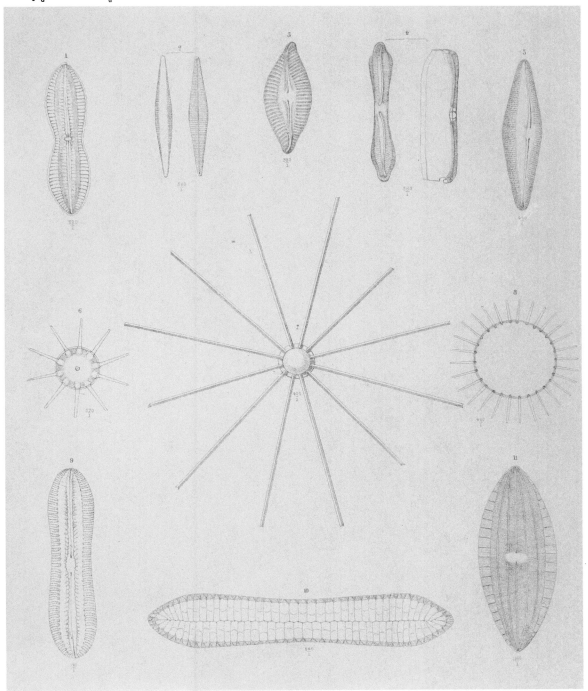

Cesare Cerri del. Castracane diresse. Micheletti inc.

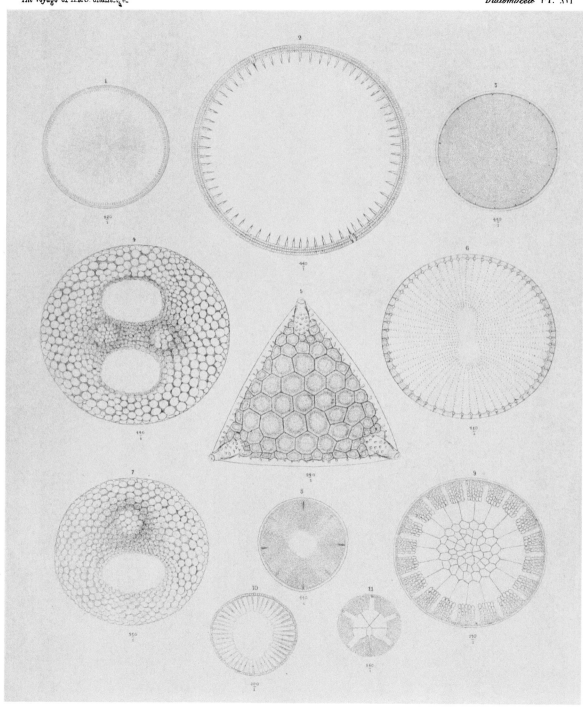

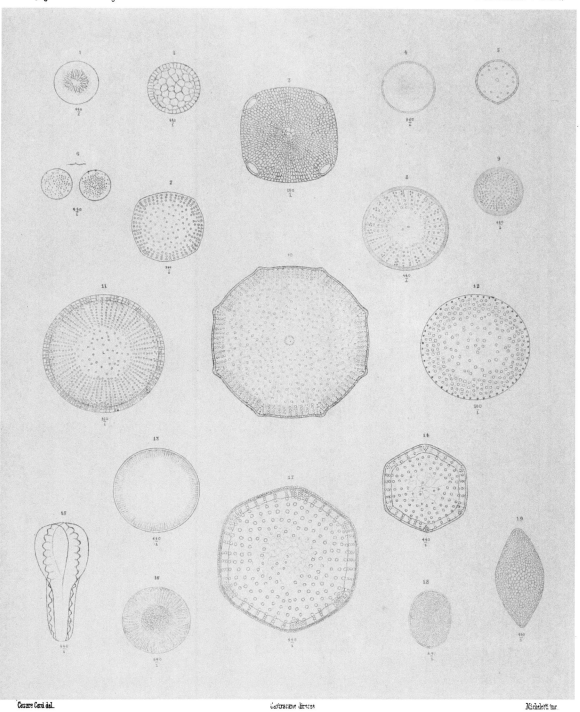

The Voyage of H.M.S. "Challenger" *Diatomaceæ* Pl. XVIII

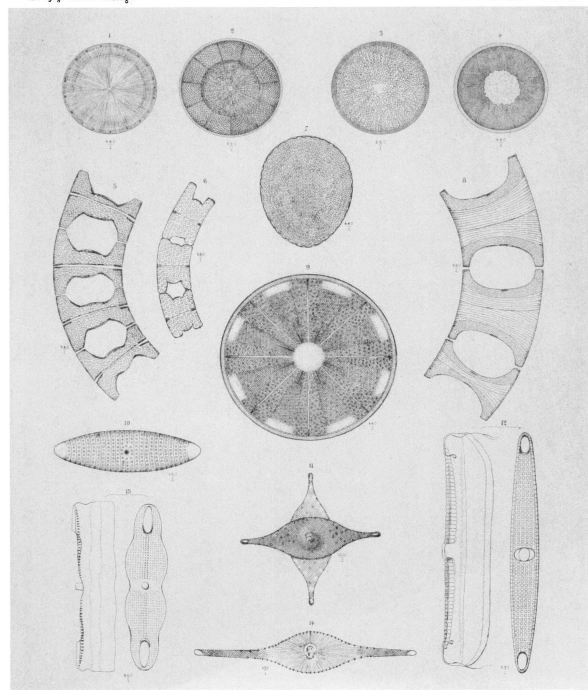

Cesare Cerù del.   Castracane diresse   Micheletti inc.

The Voyage of H.M.S. Challenger. Diatomaceæ Pl. XIX.

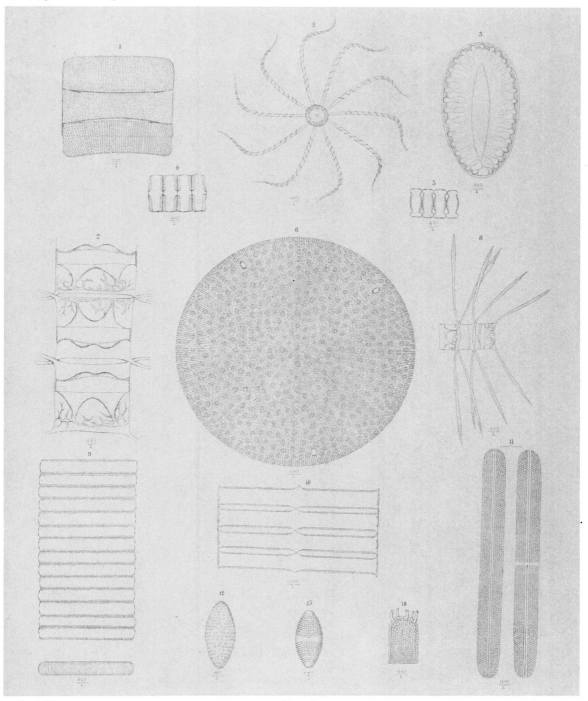

Cesare Cerù del. Castracane direxse Micheletti inc.

The Voyage of H.M.S. Challenger

Diatomaceæ Pl. XX

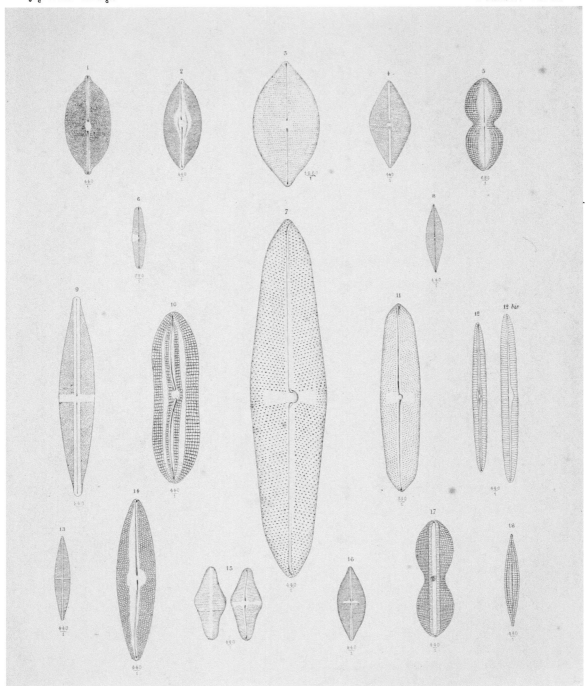

Cesare Ceru del.

Castracane diresse.

Michetti in.

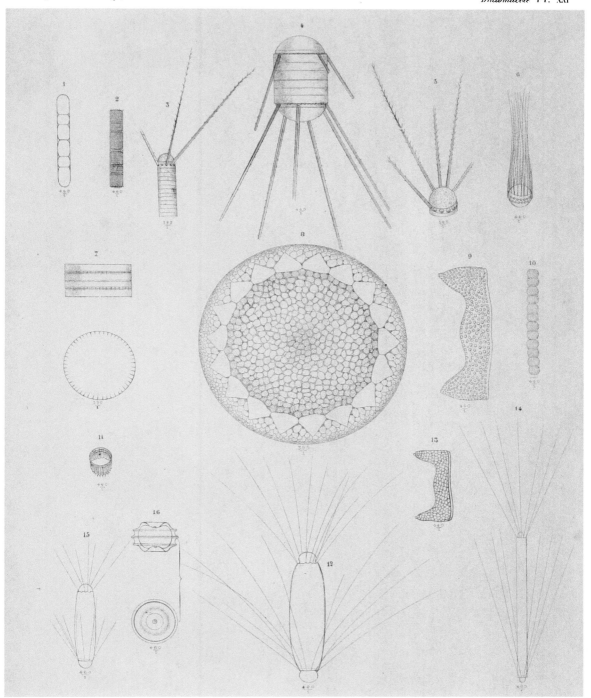

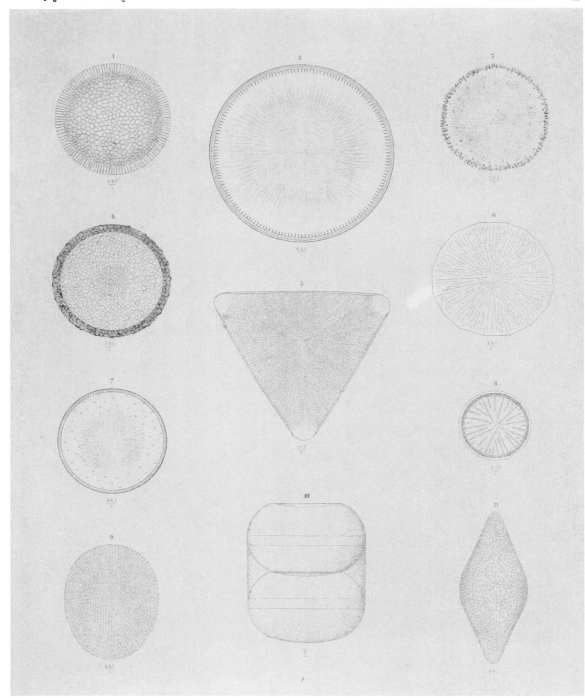

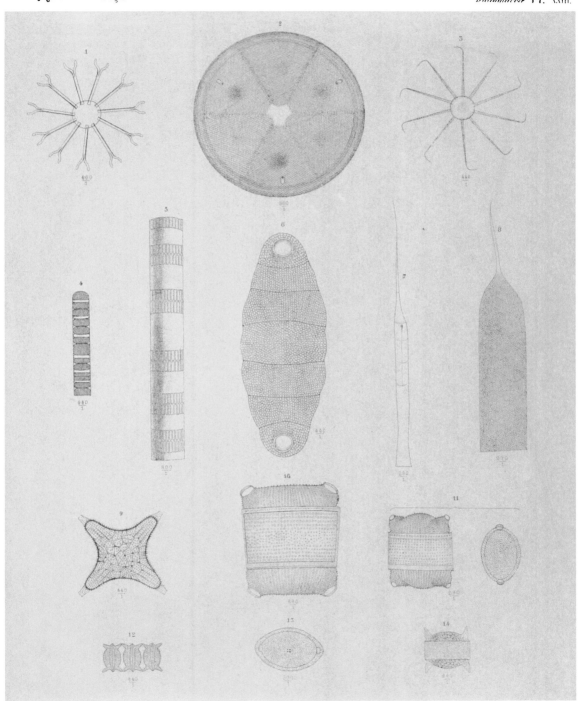

The Voyage of H.M.S. Challenger. *Diatomaceæ* Pl. XXIV.

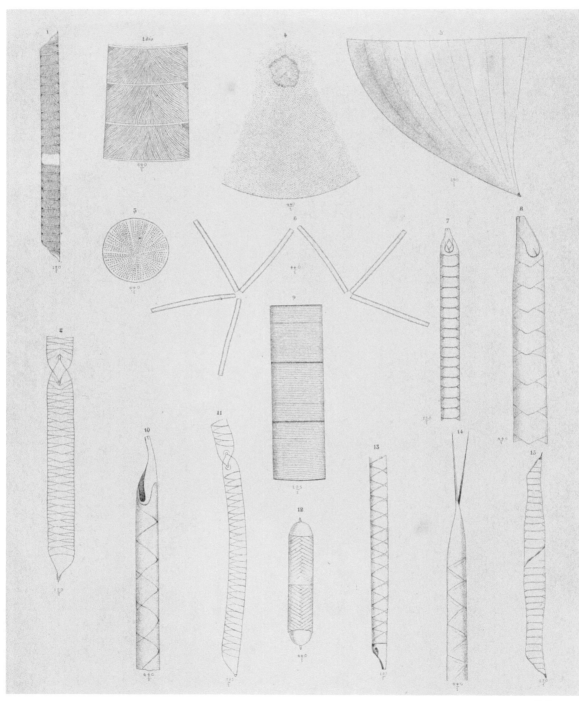

Cesare Cerri del. Castracane diresse. Micheletti imp.

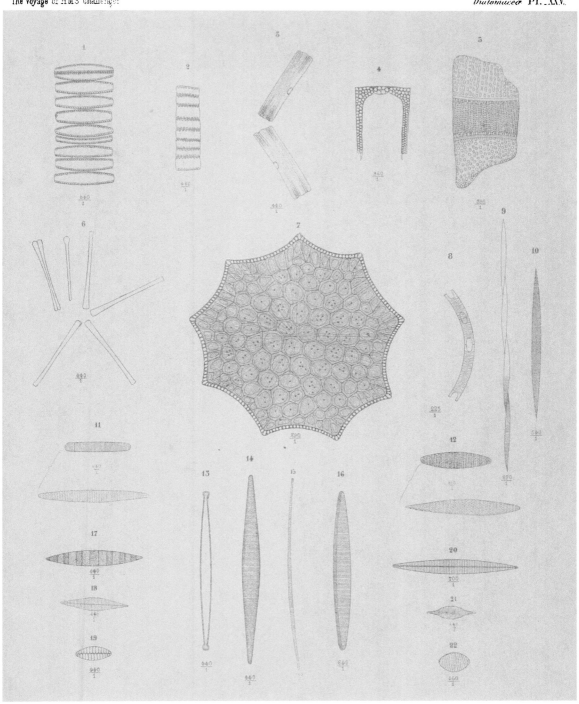

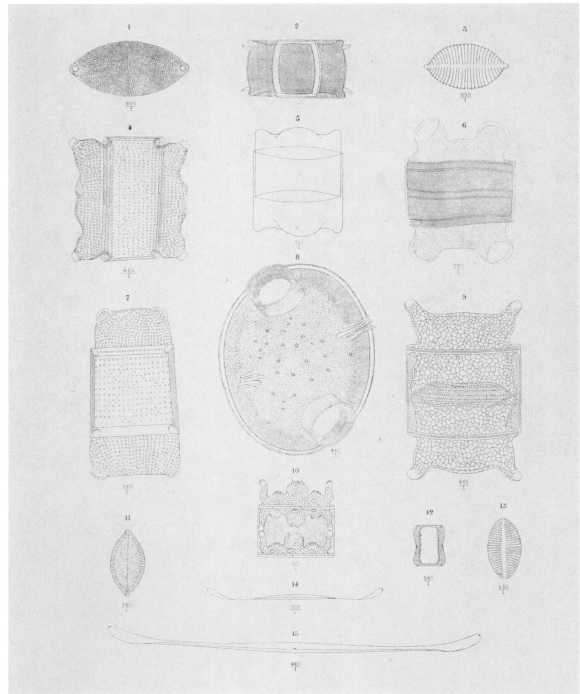

The Voyage of H.M.S. "Challenger". *Diatomaceæ.* Pl. XXVII

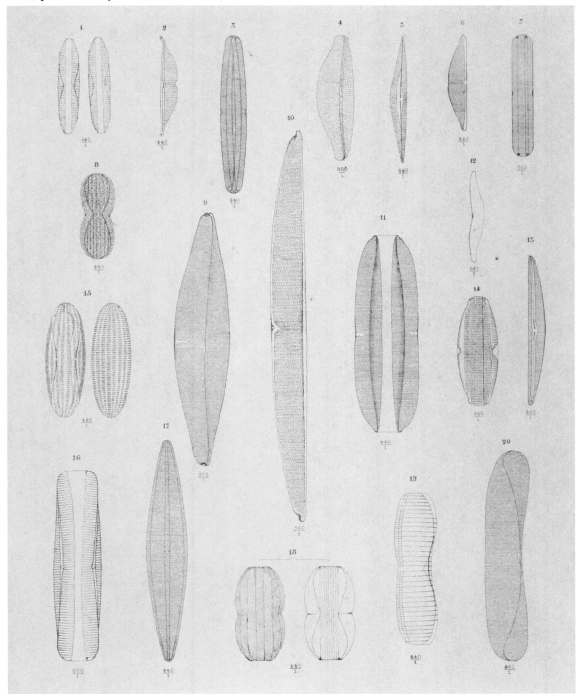

Cesare Gerli del.   Castracane diresse   Micheletti inc.

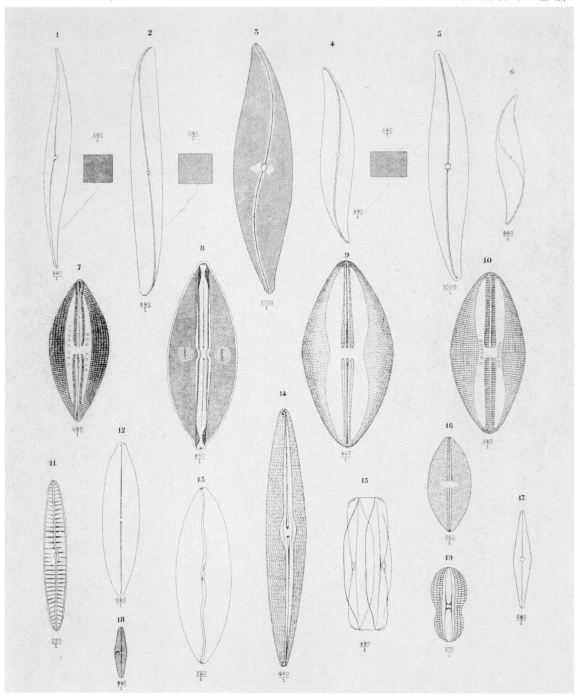

The Voyage of H.M.S. "Challenger". *Diatomaceæ* Pl. XXIX.

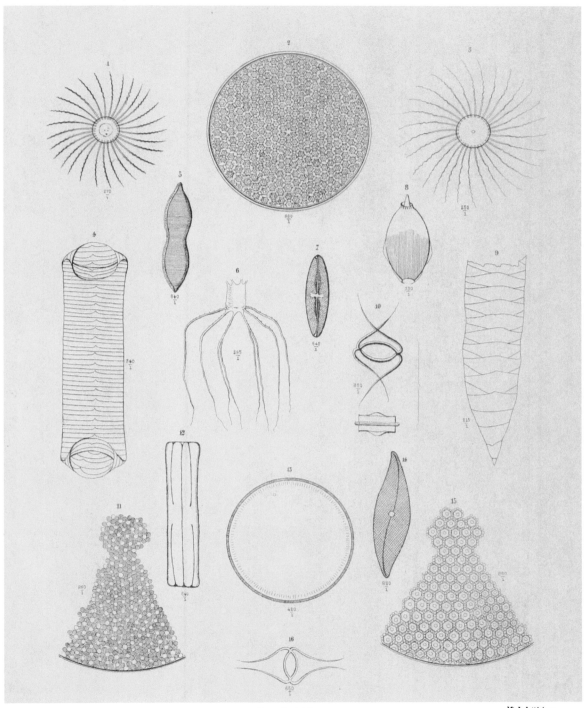

Castracane direx. Cesare Cerù del. Micheletti inc.

The Voyage of H.M.S. Challenger. Diatomaceæ Pl. XXX.

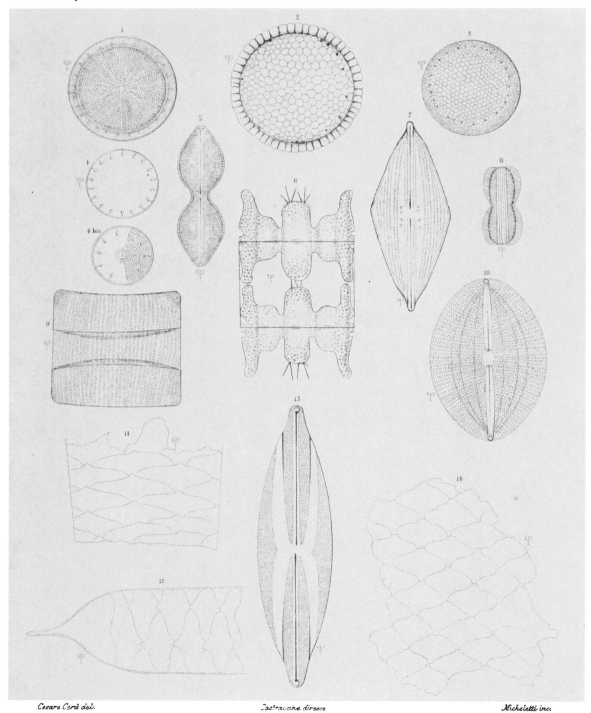

Cesare Cerù del. Castracane diresse. Micheletti inc.

第四十三卷物理化学—彩图

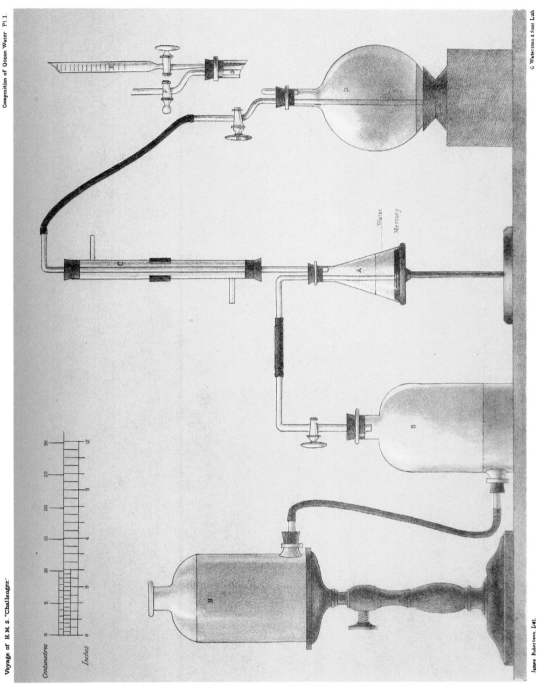

APPARATUS FOR DETERMINING THE CARBONIC ACID IN SEA WATER.

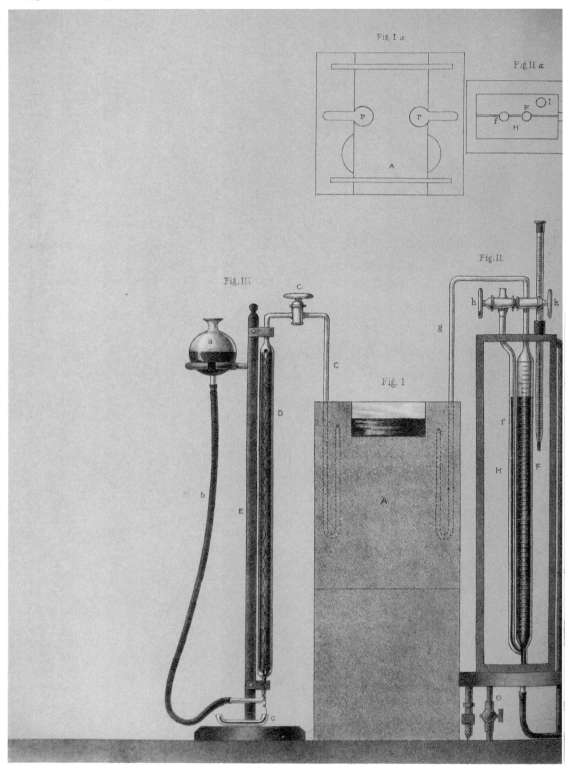

Voyage of H.M.S. "Challenger"

APPARATUS

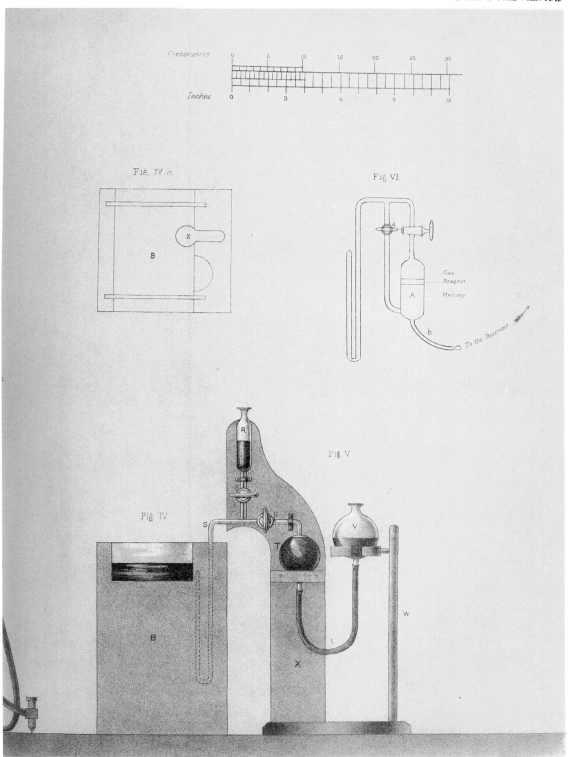

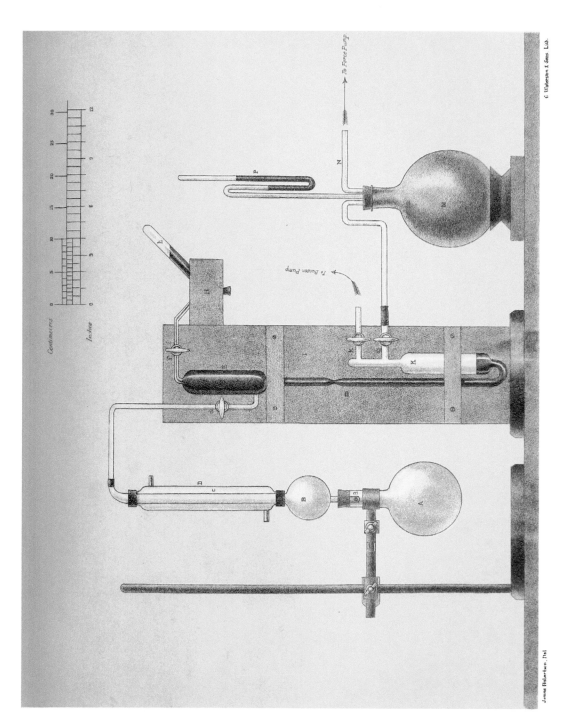

APPARATUS FOR EXTRACTING GASES FROM WATER.

第四十五卷概述一上彩图

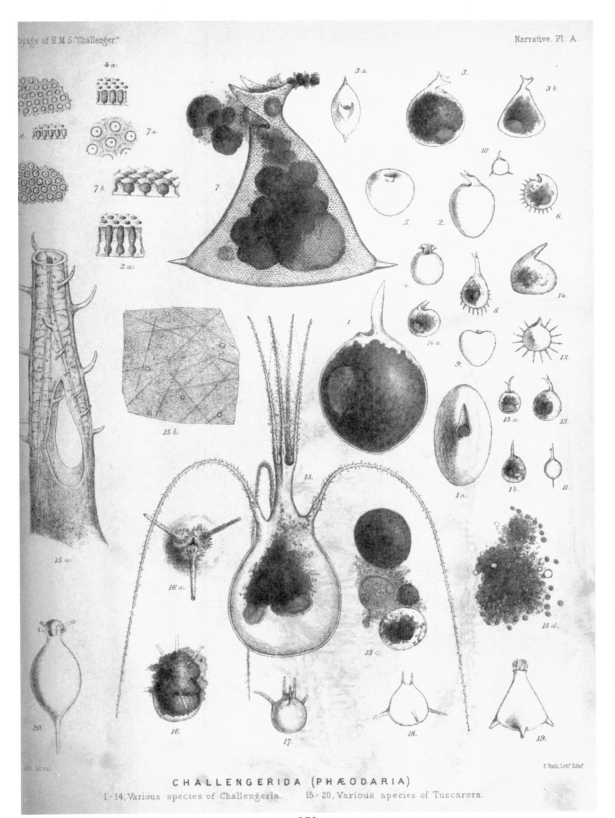

CHALLENGERIDA (PHÆODARIA)
1–14, Various species of Challengeria.   15–20, Various species of Tuscarora.

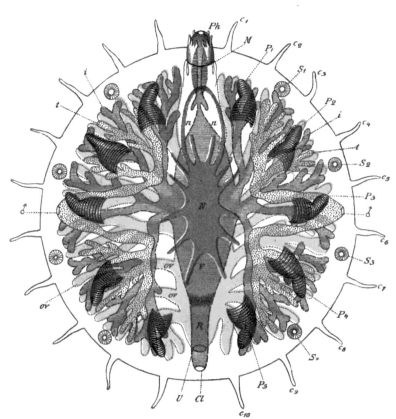

Fig. 125.—Diagram showing the Structure of *Myzostoma*.

ANTARCTIC ICEBERGS.

1-3. seen February 14th 1874, Lat 65° 42' S., Long. 79° 49' E.   4. seen February 15th 1874, Lat. 65° 59' S., Long. 78° 24' E.

ANTARCTIC ICEBERGS.

1, seen February 15th 1874, Lat. 65° 59' S., Long. 78° 24' E.   2-4, seen February 16th 1874, Lat. 66° 40' S., Long. 78° 22' E.
5, seen February 19th 1874, Lat. 64° 37' S., Long. 85° 49' E.

ANTARCTIC ICEBERGS.

1 & 2, seen February 21st 1874, Lat. 63° 30' S., Long. 89° 6' E.   3, seen February 22nd 1874, Lat. 63° 30' S., Long. 90° 47' E.
4, seen February 25th 1874, Lat. 63° 49' S., Long. 94° 51' E.

1, 2, 3 AND 4. VARIOUS DANCING COSTUMES WORN AT NAKELLO, FIJI. N° 1 AND N° 2 OF THE FISHERMAN TRIBE.
N° 5. A TONGAN TO SHOW THE COLOUR OF THE RACE.

第四十六卷概述一下彩图

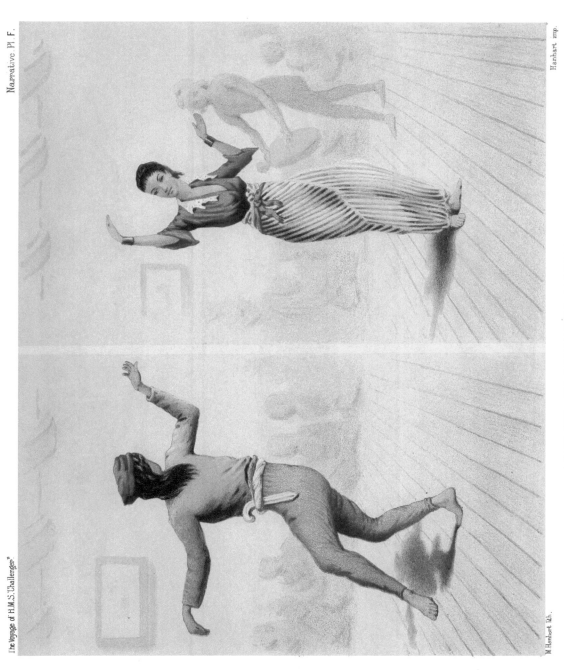

LUTAO GIRL AND BOY, DANCING, AT SAMBOANGAN.

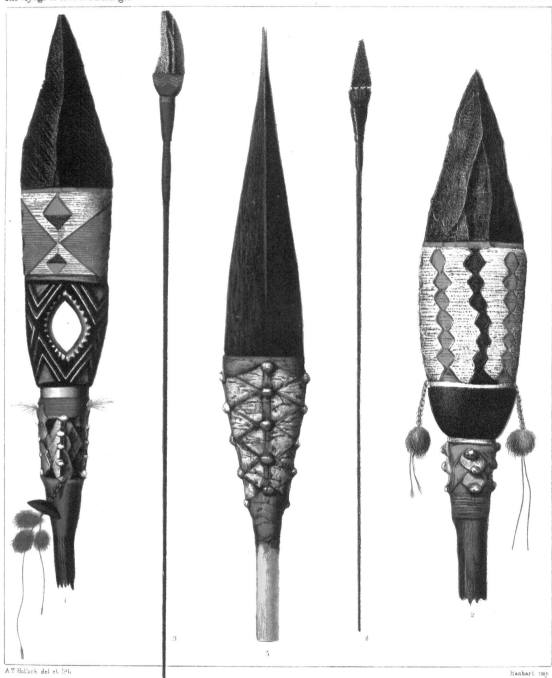

ADMIRALTY ISLANDS.

1 & 2. UNUSUALLY LARGE & HIGHLY DECORATED OBSIDIAN BLADED SPEAR. 3 & 4. OBSIDIAN BLADED SPEARS OF OLD PATTERN HEADS.
5. SPEAR HEAD WITH BLADE OF HARD WOOD PAINTED TO RESEMBLE OBSIDIAN

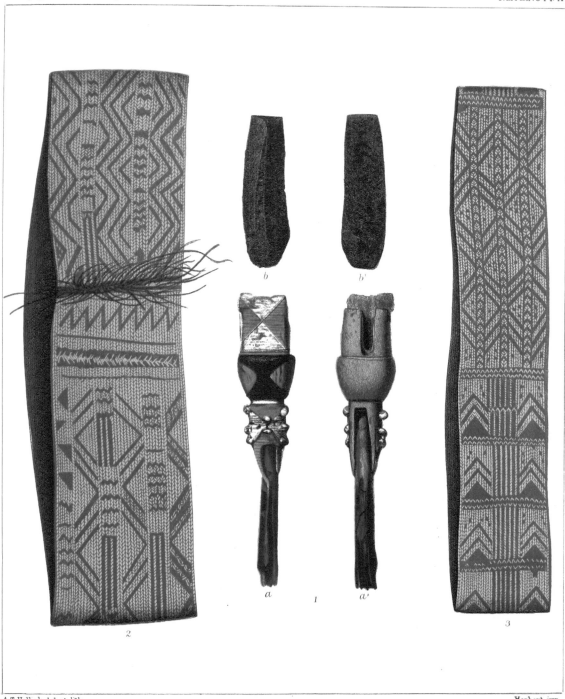

ADMIRALTY ISLANDS.

1. THE HEAD AND SHAFT OF A SPEAR WITH THE BINDING TWINE CUT OFF TO SHOW THE CONSTRUCTION. a, a'. FRONT & BACK OF THE SOCKET. b, b'. THE BLADE (BROKEN AT THE POINT.) 2, 3. TWO BELTS OF PLAITED WORK.

ADMIRALTY ISLANDS.

1. OBSIDIAN KNIFE MADE BY BREAKING OFF THE HEAD OF A SPEAR. 2. KNIFE MADE FOR THAT PURPOSE ONLY. 3. SHEATH OF BANANA LEAF FOR KNIFE OR SPEAR. 4. COMB. 5. BAG.

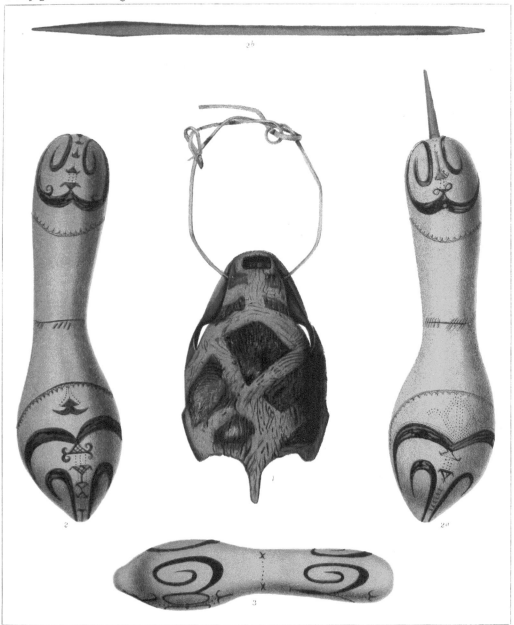

ADMIRALTY ISLANDS.

1. DECORATED SKULL OF TURTLE HUNG UP IN CLUB HOUSES AS A MEMENTO OF THE FEAST.
2a. VIEWS OF A LIME GOURD FOR CARRYING LIME TO BE CHEWED WITH BETEL. 2b STICK FOR SPOONING OUT THE LIME. 3. A SMALLER GOURD.

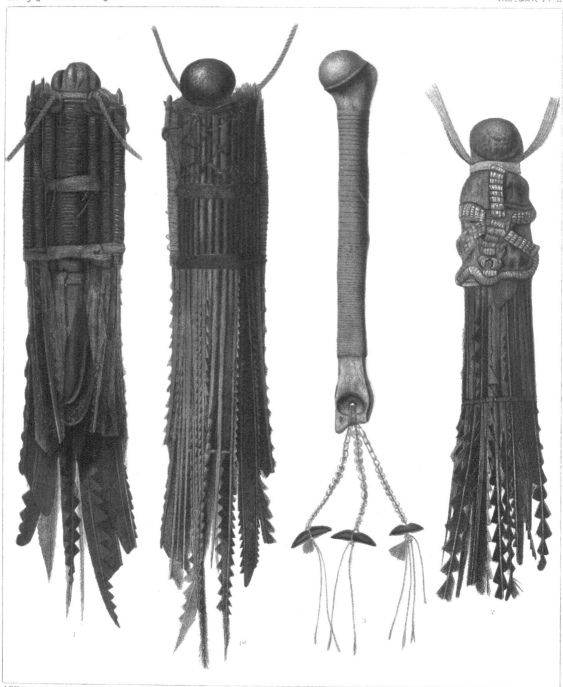

ADMIRALTY ISLANDS.

1, 1a & 2. BADGES OF DISTINCTION CONSISTING OF A HUMAN HUMERUS DECORATED WITH FEATHERS CARRIED BY WARRIORS 1, 1a FRONT AND BACK VIEWS OF SAME. 3. A SIMILAR BADGE BUT WITHOUT FEATHERS.

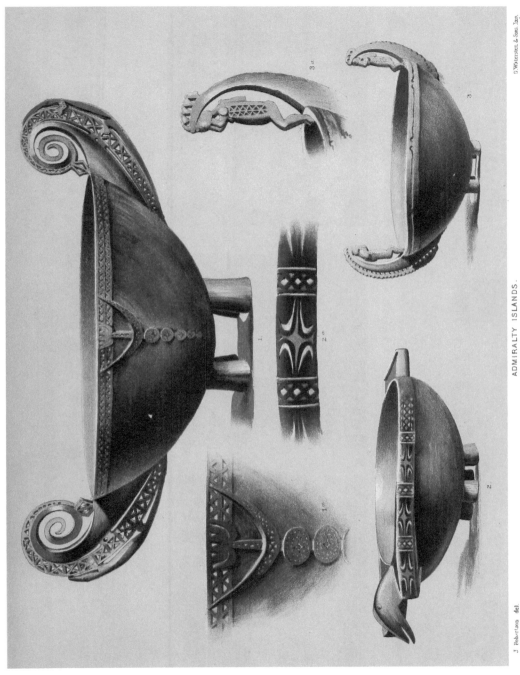

ADMIRALTY ISLANDS.

1, 1ª. A VERY FINE BOWL WITH LIZARDS ON THE HANDLES.
2, 2ª. A BOWL IN THE FORM OF A BIRD. 3, 3ª. A BOWL WITH HUMAN FIGURES SUPPORTING THE SCROLLS.
BELONGING TO J.Y. BUCHANAN Esqr.

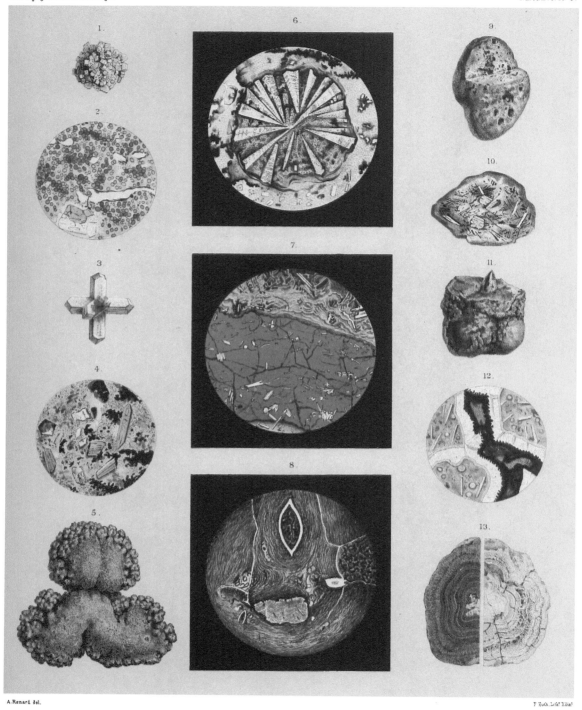

Deep Sea Deposits.

第四十七卷概述二彩图

The Voyage of H.M.S. "Challenger." — St. Paul's Rocks

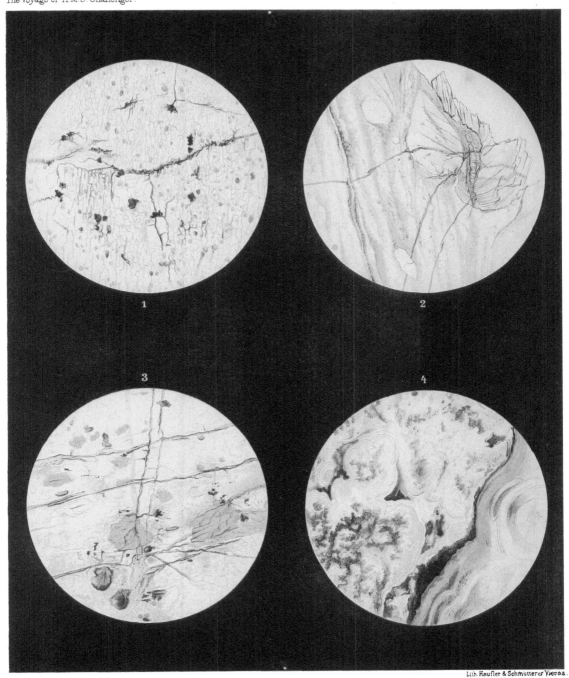

Lith. Haufler & Schmutterer, Vienna.

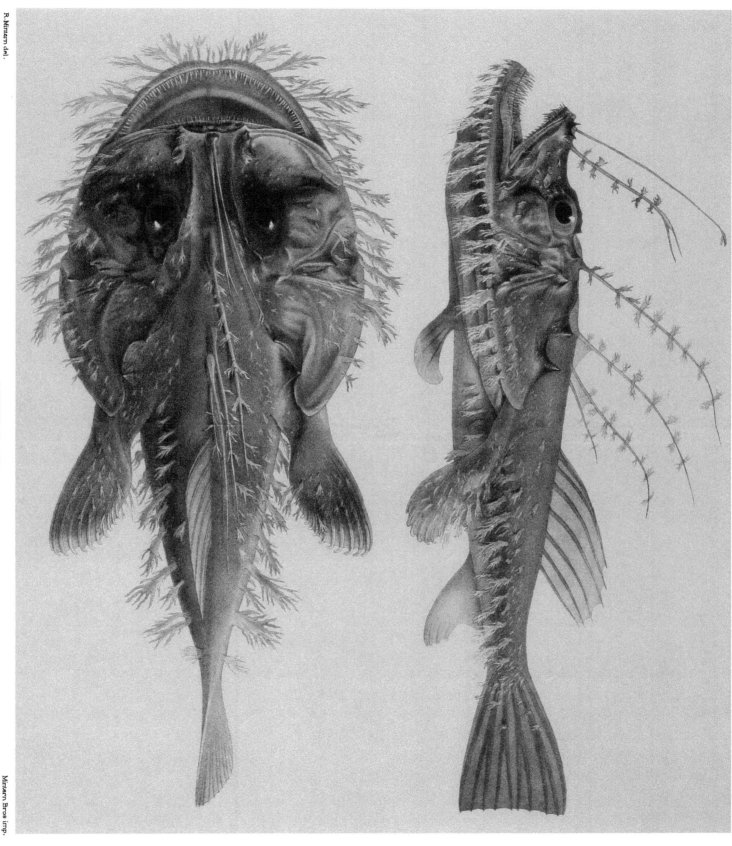

LOPHIUS NARESII.
(Admiralty Islands.)

APHRITIS GOBIO.
(Straits of Magelhaen.)